WATER RESOURCES RESEARCH
Problems and Potentials for Agriculture and Rural Communities

WATER RESOURCES RESEARCH

Problems and Potentials for Agriculture and Rural Communities

Ted L. Napier, Donald Scott
K. William Easter, and Raymond Supalla
Editors

Published for

North Central Committee 111,
Natural Resource Use and Environmental Policy

by the

Soil Conservation Society of America
7515 Northeast Ankeny Road
Ankeny, Iowa 50021

Library of Congress Catalog Card Number 83-4821

ISBN 0-935734-10-4

$6.00

Library of Congress Cataloging in Publication Data

Main entry under title:

Water Resources Research.

  Includes index.
  1. Water-supply, Agricultural—Research—Middle West.
2. Water-supply, Rural—Research—Middle West. 3. Water-supply,
Agricultural—Middle West. 4. Water-supply, Rural—Middle West.
I. Napier, Ted L. II. North Central Research Committee 111—
Natural Resource Use and Environmental Policy (U.S.).
S494.5.W3W373 1983   333.91'3   83-4821
ISBN 0-935734-10-4

SOIL CONSERVATION SOCIETY OF AMERICA

The Soil Conservation Society of America, founded in 1945, is a nonprofit scientific and educational association dedicated to advancing the science and art of good land use. Its 13,000 members worldwide include researchers, administrators, educators, planners, technicians, legislators, farmers and ranchers, and others with a profound interest in the wise use of land and related natural resources. Most academic disciplines concerned with the management of land and related natural resources are represented.

Opinions, interpretations, and conclusions expressed in this book are those of the authors.

This book is based on material presented at a symposium on water resources research, held November 9-10, 1982, at Illinois Beach State Park, Zion, Illinois. Funding for the symposium and a subsidy for publication of this book were provided by the Ford Foundation; the Farm Foundation; the Soil Conservation Service, U.S. Department of Agriculture; the Natural Resource Economics Division, Economic Research Service, U.S. Department of Agriculture; and the North Central Research Committee 111, Natural Resource Use and Environmental Policy. The contributions of these groups are gratefully acknowledged. Special recognition is given to Max Schnepf of the Soil Conservation Society of America whose editorial skills greatly improved the readability of the papers. The prompt publication of the text by SCSA is also appreciated. Without the contributions of these groups this text would not have been possible.

North Central Research Committee 111, Natural Resource Use and Environmental Policy

Dr. Arlo W. Biere
Department of Economics
Kansas State University

Dr. Richard C. Bishop
Department of Agricultural Economics
University of Wisconsin

Dr. John Branden
Department of Economics
University of Illinois

Dr. Melvin L. Cotner
Natural Resources Economics Division
Economic Research Service

Dr. Gene S. Cox
School of Forestry, Fisheries and Wildlife
University of Missouri

Dr. Robert M. Dimit, Secretary
Department of Rural Sociology
South Dakota State University

Dr. K. William Easter, Vice-chairman
Department of Agricultural and Applied
   Economics
University of Minnesota

Dr. Earl O. Heady
Center for Agricultural and Rural
   Development
Iowa State University

Dr. Thomas Hertel
Department of Agricultural Economics
Purdue University

Dr. Lee Kolmer, Administrative Advisor
College of Agriculture
Iowa State University

Dr. Jay A. Leitch
Department of Agricultural Economics
North Dakota State University

Dr. Larry Libby
Department of Agricultural Economics
Michigan State University

Dr. Ted L. Napier, Chairman
Department of Agricultural Economics
   and Rural Sociology
Ohio State University

Dr. Donald F. Scott, NCA-12
   Representative
Department of Agricultural Economics
North Dakota State University

Dr. Raymond Supalla
Department of Agricultural Economics
University of Nebraska

Contents

III. Water Research Needs and Potentials

IV. Water Research Organization and Funding Alternatives

Contributors

ARLO BIERE, Research Agricultural Economist
Kansas Agricultural Experiment Station, Kansas State University, Manhattan

ELIZABETH G. BRYANT, Research Associate
Department of Agricultural Economics and Rural Sociology, Ohio State University, Columbus

MICHAEL V. CARTER, Research Associate
Department of Agricultural Economics and Rural Sociology, Ohio State University, Columbus

LEE A. CHRISTENSEN, Agricultural Economist
Department of Agricultural Economics, University of Georgia, and Economic Research Service, U.S. Department of Agriculture, Athens, Georgia

K. WILLIAM EASTER, Professor
Department of Agricultural and Applied Economics, University of Minnesota, St. Paul

JAMES R. GILLEY, Professor
Department of Agricultural Engineering, University of Nebraska, Lincoln

ROY M. GRAY, Director of Economics
Soil Conservation Service, U.S. Department of Agriculture, Washington, D.C.

EARL O. HEADY, Director
Center for Agricultural and Rural Development, Iowa State University, Ames

R. J. HILDRETH, Managing Director
Farm Foundation, Oak Brook, Illinois

MARVIN E. JENSEN, National Research Program Leader—Water Management
Agricultural Research Service, U.S. Department of Agriculture, Fort Collins, Colorado

M. B. KIRKHAM, Associate Professor
Evapotranspiration Laboratory, Kansas State University, Manhattan

PETER F. KORSCHING, Associate Professor
Department of Sociology and Anthropology, Iowa State University, Ames

JAY A. LEITCH, Assistant Professor
Department of Agricultural Economics, North Dakota State University, Fargo

SUSAN M. MILLER, Research Technician
Nebraska Water Resources Center, University of Nebraska, Lincoln

JOHN A. MIRANOWSKI, Associate Professor
Department of Economics, Iowa State University, Ames

TED L. NAPIER, Professor
Department of Agricultural Economics and Rural Sociology, Ohio State University, Columbus

PETER J. NOWAK, Assistant Professor
Department of Sociology and Anthropology, Iowa State University, Ames

HOWARD W. OTTOSON, Executive Vice-President and Provost
University of Nebraska, Lincoln

WILLIAM L. POWERS, Director
Nebraska Water Resources Center, University of Nebraska, Lincoln

DONALD F. SCOTT, Chairman
Department of Agricultural Economics, North Dakota State University, Fargo

E. C. STEGMAN, Professor
Department of Agricultural Engineering, North Dakota State University, Fargo

RAYMOND J. SUPALLA, Associate Professor
Department of Agricultural Economics, University of Nebraska, Lincoln

EARL R. SWANSON, Professor
Department of Agricultural Economics, University of Illinois, Champaign-Urbana

JOHN F. TIMMONS, Professor
Department of Economics, Iowa State University, Ames

FREDERICK WORMAN, Graduate Research Economist
Kansas Agricultural Experiment Station, Kansas State University, Manhattan

Preface

This text is the second natural resources-oriented book sponsored by the North Central Research Committee 111, Natural Resource Use and Environmental Policy. The first book highlighted research findings in soil erosion and identified several potentially productive topics for further scientific investigation. Once the soil erosion text was completed, the NCR-111 group recognized that water resources management was closely aligned with soil erosion problems and that the group should capitalize on the knowledge base already established. Water quality, availability, and competing uses were unquestionably key issues in the North Central Region.

Subsequently, the editors of this text were commissioned to plan a national symposium to bring together some of the best thinking of interdisciplinary researchers on the problem of water quality, availability, and use in the North Central Region. It soon became evident that water resources problems were national in scope and that considerable transferability of research findings existed among regions of the United States. This realization precipitated a shift in focus from the North Central Region *per se* to a broader geographical perspective. Topics were selected that had implications for water resources use and availability in several regions.

The symposium was organized to present water resources research output in the following manner: (1) water resources research in a historical perspective, (2) existing water resources research endeavors, (3) potential contributions in water resources research, and (4) alternatives for conducting the needed research. The strategy was implemented, and this text is organized in like manner.

Chapters 1 through 4 look at water resources research in historical perspective. Research findings in selected topical areas are synthesized to demonstrate some successes and to provide insight into what has been accomplished. The authors of chapters 5 through 8 discuss research activities underway. These two sections essentially present a state of the art in selected areas of water resources research. Chapters 9 through 12 are devoted to assessments of potential contributions that water resources researchers in several disciplines can make to resolve problems mentioned previously in the text. The last chapter discusses alternative mechanisms for implementing these potential contributions.

Throughout the process of symposium planning and text development, there was an underlying desire on the part of the organizers to highlight major issues in water resources research and to emphasize the need for interdisciplinary cooperation. This orientation is reflected in the selection of specific issues for discussion rather than broad topical categories. The fact that several disciplines are represented in the text reflects the perceived need for interdisciplinary research programs. While the issues identified for discussion are not inclusive, they do represent several important areas of contributions made by water resources researchers. It should also be recognized that many other disciplines can and do make significant contributions to water resources research, but the scope of the symposium precluded elaboration of the disciplines involved.

The editors and contributors to this project recognize that other legitimate perspectives and problem definitions exist in water resources research. We do believe, however, that the materials presented in this text reflect the state of the art in the areas discussed and that the recommendations for future research have merit in the context of the materials presented.

Ted L. Napier
Ohio State University

April 1983

I
History of Water Resources Research

1

Water Research in the North Central Region: A Survey, 1950-1980

Howard W. Ottoson, William L. Powers and Susan M. Miller

We determined the incidence of water research related to agriculture by reviewing (1) publication and project lists from agricultural experiment stations in the North Central Region, (2) publication lists supplied by state water research institutes, and (3) computerized records of Cooperative State Research Service reports on active projects. Because an examination of all publications between 1950 and 1980 would have been too time consuming, we reviewed only those publications reported every fifth year beginning with 1950. This sample, we assumed, would reflect trends in the research.

Our examination of these trends showed that research activity began at a rather low level in 1950 and peaked in the 1970s. What factors influenced these trends? We suggest they included (1) state and federal legislation, (2) changes in the mission of federal and state agencies, (3) public opinion and awareness, and (4) advances in technology.

Federal water pollution control legislation in particular has guided research trends. The first Water Pollution Control Act (P.L. 80-845) was passed in 1948. Although amendments were added in 1956 and again in 1961, it was not until the mid-1960s and early 1970s that amendments to the act provided enough money and control to significantly influence the pace of water quality research.

Two pieces of important national legislation are the Clean Water Act of 1965 (P.L. 89-234), and the 1972 Federal Water Pollution Control Act Amendments (P.L. 92-500). The 1965 legislation established a Federal Water Pollution Control Administration within the U.S. Department of

Housing, Education, and Welfare and set water quality standards. Section 208 of the 1972 legislation had the greatest effect on agricultural research because it stressed the control of pollution from diffuse sources, including runoff from agricultural land, and created the need for data on water quality and pollution control measures.

The Water Resources Research Act of 1964 (P.L. 88-379) also influenced research through its establishment of water resources research institutes at the land grant universities in each state. The National Water Commission was created in 1968.

The National Environmental Policy Act of 1969 established the Council on Environmental Quality. About the same time the Environmental Protection Agency was created. Substantial funding became available through the two agencies for both in-house and contract research. By 1975, state and federal planning activities had increased significantly.

Public awareness of water quality issues during the late 1960s and 1970s put pressure on federal agencies to fund this type of research. State agencies felt the pressure also and expanded their research efforts.

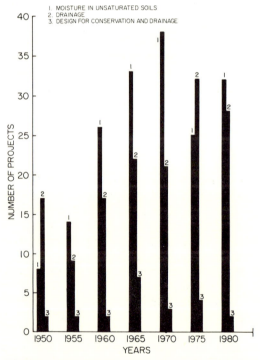

Figure 1. Frequency of projects on soil-water relationships, North Central Region, 1950-1980.

Still another factor affecting research trends was the declining water tables in the High Plains. In the 1970s it became evident in Kansas and Nebraska that groundwater supplies were being depleted. Research on groundwater recharge and conservation was needed. The resulting establishment of groundwater management districts created a demand for knowledge on irrigation efficiency.

The development of new irrigation technology had a profound effect on the quantity and type of research. The center pivot irrigation system enabled the irrigation of land previously classified as nonirrigable because of land leveling costs. In addition to making it possible to irrigate rolling land, the center pivot system saved large amounts of labor. Its adoption did much to increase water use in the western Great Plains.

Sputnik and the resulting space-age miniaturization probably did more to alter water-related research in agriculture than any other single factor. Sophisticated electronics permit the collection and storage of large quantities of data needed to help explain rapidly changing plant-water and soil-water relationships. Scanners and loggers can collect and store data several times a minute over 24-hour periods. Computers can analyze complex mathematical relationships that previously could be calculated only by slow mechanical calculators and tables of mathematical functions. This ability to collect, store, and analyze data has enabled scientists to examine dynamic water relationships in agriculture that were unheard of 20 years ago.

Physical Relationships of Water, 1950-1980

In examining the publication lists from the North Central experiment stations and the state water resources research institutes, we found in excess of 1,000 projects on the physical relationships of water related to agriculture. The sample disclosed about 50 projects in 1950. The number increased to between 200 and 225 in 1970, 1975, and 1980.

We classified the projects into six rather broad categories: (1) soil-water relationships, (2) plant-water relationships, (3) atmospheric-water relationships, (4) combined plant-soil-atmospheric water relationships, (5) hydrology, and (6) irrigation.

Soil-Water Relationships

By the late 1960s and the 1970s, research activity on soil-water relationships had more than doubled that in 1950 (Figure 1). But that activity also tended to level off during the last 15 to 20 years of our study period.

Research in this category seemed to fall rather naturally into three main groups: (1) soil moisture in unsaturated soils, (2) surface and subsurface drainage, and (3) engineering design for drainage and soil conservation.

Plant-Water Relationships

Research activity in this category was diverse and difficult to divide into homogeneous groups. With more than 200 projects in this category, however, some groupings were necessary to deal with the subject. The groups we finally developed were as follows: (1) information on plant water use, (2) water availability and plant growth, (3) interrelationships

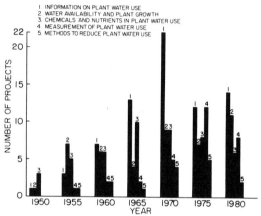

Figure 2. Frequency of projects on plant-water relationships, North Central Region, 1950-1980.

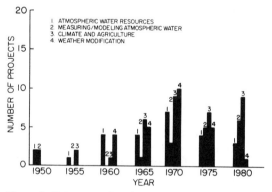

Figure 3. Frequency of projects on atmospheric-water relationships, North Central Region, 1950-1980.

of chemicals and nutrients with plant water use, (4) measurement of plant water use, and (5) methods to reduce plant water use. Figure 2 shows the trends in this category.

Atmospheric-Water Relationships

Research in this category depicts the interest in atmospheric water and its role in agriculture. We placed the projects in four major groups: (1) atmospheric water resources; (2) methods of measuring, predicting, or modeling atmospheric moisture and evaporation; (3) relationships between climate and agriculture; and (4) climate modification.

There were more than 100 publications in this category. Activity peaked at just under 30 projects in 1970 (Figure 3).

Plant-Soil-Atmospheric-Water Relationships

A number of research topics spanned two or more of the first three categories. We placed these studies, of a multidisciplinary nature, in a separate category, and made no groupings within the category. Figure 4 illustrates the frequency of activity in this category. Major interest was found in South Dakota, Nebraska, and Kansas.

Hydrology

Studies in hydrology did not clearly fall into the categories of agricultural water use. They tended to be research on basic concepts more of interest to civil engineers. Examination of the numbers of projects in this category requires caution because much of the hydrologic work is not done in the agricultural experiment stations; therefore, the reference sources we used in our survey did not include much of the work done by other units in the universities over the years. In spite of this shortcoming, we tabulated the available information because hydrologic concepts are an important factor in the management of water resources.

We divided research in this category into the following four groups: (1) availability and use of surface water and groundwater, (2) hydrologic studies on surface water and groundwater; (3) mathematical models and measurement of surface water and groundwater, and (4) engineering principles for control of water supplies (Figure 5).

Irrigation

The final category of research projects concerned irrigation practices. We placed this research in four groups: (1) technical feasibility, including methods and practices; (2) engineering principles of irrigation and equip-

ment design; (3) maximizing crop yields using chemicals; and (4) irrigation management for efficient water use (Figure 6).

Summary of Physical Relationships

Figure 7 summarizes activity in the six categories of research on physical relationships of water. In general, there was about a fourfold increase

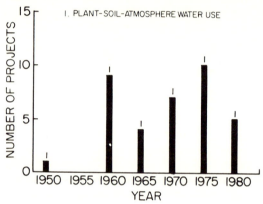

Figure 4. Frequency of projects on plant-soil-atmos-
pheric-water relationships, North Central Region,
1950-1980.

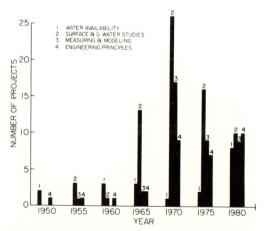

Figure 5. Frequency of projects on hydrology, North
Central Region, 1950-1980.

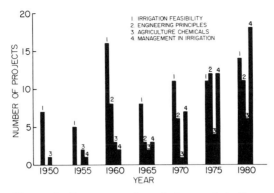

Figure 6. Frequency of projects on irrigation, 1950-1980.

in research activity between 1950 and 1970, when the activity peaked. Research in soil-water relationships roughly doubled in that time, while hydrologic studies increased nearly 13 times.

Current Research on Physical Relationships of Water

Table 1 summarizes current research projects. Half of the research on this topic is being done in Michigan, Missouri, Nebraska, and North Dakota.

Water Quality Studies, 1950-1980

A large body of literature has emerged concerning water quality in recent years. Our focus here is on agricultural-related studies.

Research activity with respect to water quality in agriculture was minimal in the North Central Region until the 1970s (Figure 8). With the advent of that decade, however, research on water quality in agriculture increased dramatically.

We arbitrarily categorized the water quality studies reviewed as follows: (1) effects of agricultural runoff and leachates on surface water and groundwater quality; (2) effects of using municipal, feedlot, or industrial wastes on cropland; and (3) saline water and agriculture.

Effects of Agricultural Runoff and Leachates

Research activity in this area was light in the North Central Region during the 1950s and 1960s (Figure 9). In this first category, we identified several groups: (1) occurrence of chemicals, organic waste, and sediment in groundwater and surface water; (2) methods of detecting and measuring pollutants in agriculture; (3) effects of pollutants on man, livestock, or aquatic ecology; and (4) studies of management or structural controls to reduce runoff or leaching. In addition, the first and fourth groups were subdivided to examine the research activity in more detail.

Occurrence of Chemicals, Organic Waste, and Sediment. We identified three subgroups, including pollution from livestock operations, pollution from crop operations, and pollution from nonspecific sources. Figure 10 shows the frequencies of studies in these subgroups.

Methods of Detecting and Measuring Pollutants. The variety of modeling effort was impressive, including such interests as: nutrient and pesticide transport, strategies for implementation of water quality management, and sedimentation in reservoirs or streams. This category also included methodological studies on preservation of water samples, the use of various indicators to determine water quality and remote sensing techniques to determine water quality.

Pollutant Effects on Man, Livestock, or Aquatic Ecology. The fate of pesticides in aquatic environments and the effects of erosion and sedimentation on fish populations are examples of research effort in this cat-

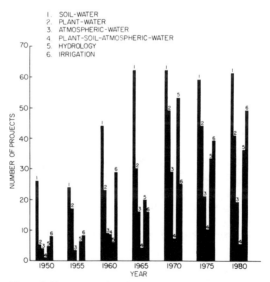

Figure 7. Frequency of projects on the physical relationships of water by major subject category, North Central Region, 1950-1980.

Table 1. Current research projects on physical relationships of water in the North Central Region.

Topic	Total	IL	IN	IA	KS	MI	MN	MO	NE	ND	OH	SD	WI
							State						
Soil-water relationships	75	3	11	9	2	2	11	9	6	3	11	4	4
1. Soil moisture in unsaturated soils	29	2	3	4	2	1	1	-	4	3	2	4	3
2. Surface and sub-surface drainage	36	1	8	5	-	-	8	7	1	-	5	-	1
3. Engineer design for drainage and conservation	8	-	-	-	-	1	2	-	1	-	4	-	-
Plant-water relationships													
1. Information on plant water use	32	-	2	4	4	1	3	1	9	4	2	-	2
2. Water availability and plant water use	6	-	-	-	1	-	-	-	2	-	2	-	1
3. Chemicals, nutrients, and plant water use	9	-	1	3	1	-	-	-	-	3	-	-	1
4. Measurement of plant water use	4	-	-	-	-	1	-	1	1	1	-	-	-
5. Methods to reduce plant water use	4	-	-	1	1	-	-	-	2	-	-	-	-
Atmospheric-water relationships	-	-	-	-	-	-	-	-	-	-	-	-	-
Plant-soil-atmospheric-water relationships	12	-	2	2	4	-	-	-	1	1	1	1	-
Hydrology	17	-	2	-	4	2	2	5	-	1	1	-	-
1. Availability use of surface water and groundwater	3	-	1	-	1	1	-	-	-	-	-	-	-
2. Hydrologic studies on surface water and groundwater	8	-	-	-	2	1	2	2	-	1	-	-	-
3. Mathematical models and measurement of surface water and groundwater	5	-	1	-	-	-	-	3	-	-	-	1	-
4. Engineering principles for control of water supplies	1	-	-	-	1	-	-	-	-	-	-	-	-
Irrigation	53	4	-	-	11	1	3	1	13	11	1	5	3
1. Technical feasibility, methods, and practices	18	-	-	-	5	-	1	-	2	6	1	2	1
2. Engineering principles of irrigation and equipment design	14	-	-	-	4	-	-	-	4	3	-	3	-
3. Maximizing crop yields using agricultural chemicals	4	-	-	-	-	-	-	-	2	2	-	-	-
4. Irrigation management for efficient water use	17	4	-	-	2	1	2	1	5	-	-	-	2
Total	355	14	31	28	45	12	35	30	53	39	30	20	18

egory. Also of interest were the role of phosphorous and nitrates in surface water supplies and nitrates in groundwater. Several studies were concerned with the effects of various chemicals on the physical properties of water.

Management or Structural Controls.   We subdivided the projects reviewed in this group into those relating to livestock management and those relating to crop production management (Figure 11). A North Central Region project, "Animal Waste Management with Pollution Control" (NC-93), involved several states.

Effects of Using Wastes on Cropland.   Seven of the North Central States had projects investigating various aspects of using livestock wastes in crop production during the 1970s. Kansas focused heavily on the use

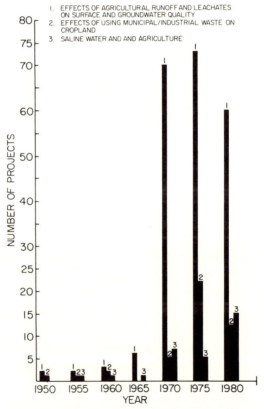

Figure 8. Frequency of projects on water quality related to agriculture, North Central Region, 1950-1980.

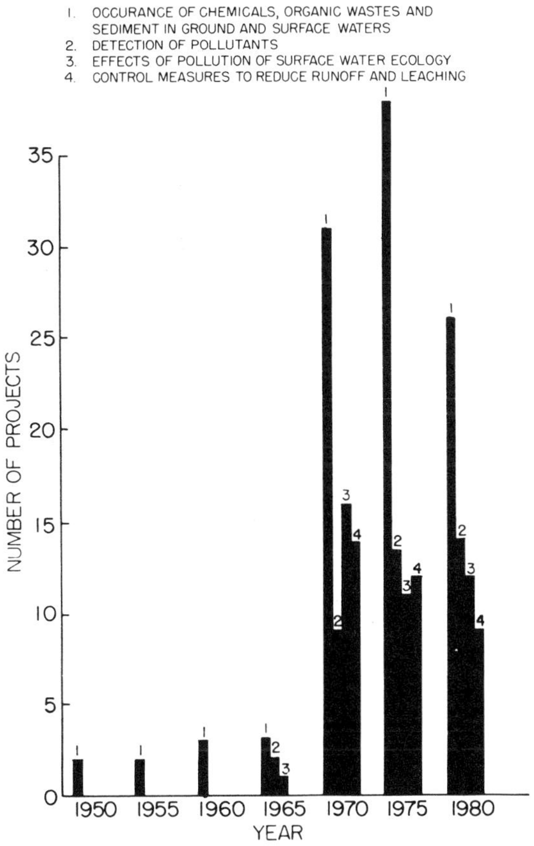

Figure 9. Frequency of projects on agricultural runoff and water quality, North Central Region, 1950-1980.

of feedlot wastes. Indiana and Wisconsin studied the use of dairy wastes while Iowa emphasized swine wastes.

Saline Water and Agriculture. Research in this category included detection of saline seeps using remote sensing and management of saline seeps through cropping, tillage, and irrigation practices. Salt pollution of groundwater and salt stress in plants were also topics of interest.

Current Research on Water Quality

Although the current research information system of the U.S. Department of Agriculture (USDA) and the state agricultural experiment sta-

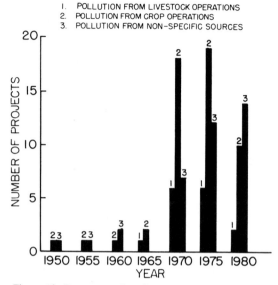

Figure 10. Frequency of projects on the occurrence of chemicals, organic wastes, and sediments in ground and surface waters.

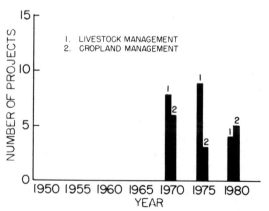

Figure 11. Frequency of projects on control measures to reduce runoff and leaching, North Central Region, 1950-1980.

Table 2. Current research projects on water quality in the North Central Region.

Topic	Total	IL	IN	IA	KS	MI	MN	MO	NE	ND	OH	SD	WI
Agricultural runoff and water quality													
1. Chemicals, organic wastes, and sediments in general and surface waters	44	4	4	7	8	1	4	5	-	2	5	1	3
2. Methods of determining, measuring, predicting, or modeling water pollutants	7	-	-	-	1	-	2	1	1	-	-	-	2
3. Effects of pollutants on surface water ecology or use by man and animals	16	2	-	1	2	3	-	1	2	-	2	1	2
4. Management or structural controls to reduce runoff or leaching	11	-	-	1	1	-	2	-	2	-	2	1	2
Use of municipal, livestock, or industrial wastes for crop production	12	-	1	1	3	2	2	-	-	-	2	-	1
Saline water and soils and sodic soils and water quality	8	-	-	-	1	-	-	-	-	7	-	-	-
Chemical changes in atmospheric deposition and its effect on agricultural land and surface water	7	1	1	1	-	1	-	-	1	-	1	-	1
Total	105	7	6	11	16	7	10	7	6	9	12	3	11

tions does not include all research on water quality taking place in the North Central States, it does include an important portion of all such research in the region. Our examination of the active projects revealed substantial activity (Table 2).

Water Economics and Management, 1950-1980

In this section of our historical review, we examined the following research areas: (1) economic aspects of water utilization in agriculture, (2) legal aspects, (3) planning or policy aspects of water utilization, and (4) management of water resources, including institutional framework. Our sampling of the literature indicated that a fairly substantial research effort on the economics of water resources in agriculture was underway during the entire postwar period (Figure 12). There was an increase in activity between 1950 and 1960. After 1960, the level of activity held rather constant. Legal studies in the area of water began during the 1960s, increased dramatically by 1970, then continued at a somewhat reduced level during the 1970s. Planning and policy studies as well as implementation studies also increased sharply by 1970 and remained at a high level of activity throughout that decade.

Economic Aspects of Water Utilization

With respect to the economic impacts of water use, we identified five general research groups: (1) costs and benefits of irrigation to individual operators, (2) economics of development and control, (3) economics of erosion and erosion control, (4) economic impacts of maintaining or improving water quality, and (5) economic impacts associated with weather and weather control.

Figure 13 summarizes the incidence of projects among these components of the economics of water utilization. Interestingly, the four Great Plains States in the North Central Region accounted for nearly half of the economic studies on water in agriculture during this period.

Economics of Erosion and Erosion Control. This topical area addressed the cost of soil erosion in terms of both reduced productivity associated with soil loss and the cost and benefits of erosion control measures. Our sample of research in the North Central Region identified 27 projects in six states during the postwar period. Obviously, a much larger

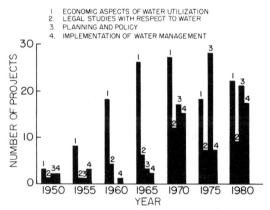

Figure 12. Frequency of projects on economics and management of water resources, North Central Region, 1950-1980.

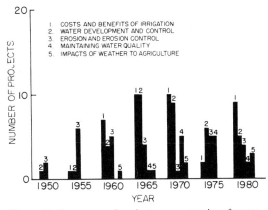

Figure 13. Frequency of projects on economics of water utilization, North Central Region, 1950-1980.

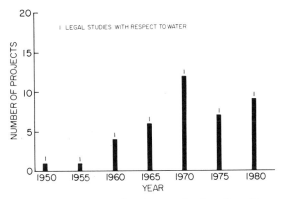

Figure 14. Frequency of projects on legal studies with respect to water, North Central Region, 1950-1980.

number of projects could have been identified if we had broadened our investigation to include more general aspects of conservation farming. The projects that we identified were more narrowly defined to include those related specifically to soil erosion as the result of the effects of water, the related costs of erosion, and measures that could be used to reduce soil loss.

Economics of Maintaining or Improving Water Quality. The studies in this group were largely concerned with the control of organic and inorganic chemicals. We included research on the cost of controlling sedimentation in the preceding section.

Most research on the economics of water quality occurred during the 1970s. Thirteen projects were carried on in eight North Central States. Despite the newness of this field of research, the variety of water pollution topics that were the target of economic analysis was impressive.

Legal Aspects

The 1960s saw the emergence of legal analysis of issues relating to water in agriculture. Figure 14 indicates the volume of activity in the region since that time.

Planning or Policy Aspects

We included in this category investigations related to (1) water policies and management procedures; (2) decision-making methodology; (3) data acquisition and handling, including analysis for decision-making; and (4) public attitudes and citizen involvement with respect to water policy. Investigations in this area were few in number until 1970 (Figure 15). During the 1970s, however, most North Central States became involved in research on these topics.

Management of Water Resources

The four groupings in this category included projects relating to (1) organizational arrangements, (2) taxing or pricing mechanisms, (3) water development funding, and (4) socioeconomic considerations in developing water resources. Figure 16 displays the activity, as reflected in our sample projects since World War II.

Current Research on Water Economics and Management

Table 3 shows the overall status of current research by the state agricultural experiment stations and USDA. Half of the work in the North Central Region on management of water resources involves the economics of water utilization. Water planning and policy account for another fourth of the projects, with management of water resources accounting for most of the remainder. A surprisingly small effort in the region is devoted to legal studies with respect to water.

Summary

The numbers of projects identified in our survey indicate that water research in the North Central Region has escalated fivefold over the past 30 years. While some of this increased activity may be due to improvements in project reporting procedures over the years, the data through 1965 suggest a gradual increase in water-related research. Rather sharp increases were noted between 1965 and 1970. The general trend since 1970 has been a very slight decline in total research projects.

Throughout the period, the general area of water economics and man-

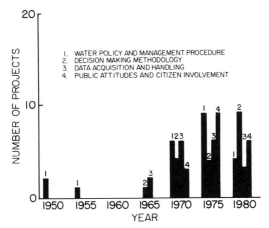

Figure 15. Frequency of projects on water planning and policy, North Central Region, 1950-1980.

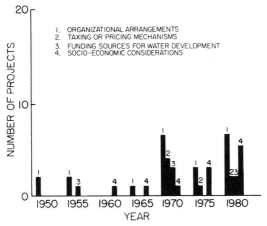

Figure 16. Frequency of projects on implementation of water management, North Central Region, 1950-1980.

agement has maintained a fairly stable portion of the total research effort, averaging 17 percent. Nearly twice as many projects were reported in 1970 as in 1965, but only gradual increases occurred in the preceding years, followed by a generally stable number of projects after 1970.

For the first 20 years, water quality studies accounted for about 5 percent of the total research identified by our survey. In 1970, this research area showed a 10-fold jump in reported projects, representing 22 percent

Table 3. *Current research projects on management of water resources in the North Central Region.*

Topic	Total	IL	IN	IA	KS	MI	MN	MO	NE	ND	OH	SD	WI
Economics of water utilization													
1. Costs and benefits of irrigation	9	1	-	-	5	-	-	-	-	-	-	2	1
2. Economics of water development and control	8	-	1	1	3	-	-	1	1	1	-	-	-
3. Economics of erosion and erosion control	4	1	-	1	-	-	-	-	-	-	-	1	1
4. Economics of water quality	3	-	-	-	1	-	-	-	-	-	1	-	1
5. Economic impact of weather in agriculture	-	-	-	-	-	-	-	-	-	-	-	-	-
Legal studies with respect to water	1	-	-	-	-	-	-	-	1	-	-	-	-
Water planning and policy													
1. Water policies and management procedure	9	1	-	1	2	2	-	-	1	-	-	1	1
2. Decision-making methodology	-	-	-	-	-	-	-	-	-	-	-	-	-
3. Data acquisition and handling	1	-	-	-	-	-	-	-	1	-	-	-	-
4. Public attitude and involvement	-	-	-	-	-	-	-	-	-	-	-	-	-
Implementation of water management													
1. Organizational arrangements	1	-	-	1	-	-	-	-	-	-	-	-	-
2. Taxing or pricing mechanism	1	-	-	-	-	-	1	-	-	-	-	-	-
3. Funding source	-	-	-	-	-	-	-	-	-	-	-	-	-
4. Socioeconomic considerations in water policy implementation	6	-	1	3	1	-	-	-	-	-	1	-	-
Total	43	3	2	7	12	2	1	1	4	1	2	4	4

of the total research effort. It has maintained an average of 24 percent for the period since 1970. Research projects in this area actually peaked in 1975, then fell off somewhat in 1980.

Physical relationships of water represented the greatest research effort over the 30-year period, claiming 78 percent of the total during the first 15 years and 57 percent during the last 10. A significant increase appeared between reporting years 1955 and 1960 with twice as many projects counted in 1960. A doubling again occurred between sample years 1965 and 1970. Sample years after 1970 showed relatively little change.

We should note that our survey revealed a small amount of research done by the agricultural experiment stations on various aspects of wetlands. Three active projects on wetlands were reported by CRIS (Current Research Information Service). We expect that wetlands research projects would have occurred more frequently if other reference sources had been sampled. However, the appearance of wetlands research in more recent experiment station literature may indicate an increasing interest in this subject by the agricultural colleges.

2

Irrigation Management: Contributions to Agricultural Productivity

James R. Gilley and Marvin E. Jensen

Water is one of the most limiting of natural resources affecting agricultural crop production. In humid parts of the North Central Region, excess water often delays planting in the spring, damages established crops, and may prevent or delay harvest. Short periods of drought in more humid areas may reduce crop yields, especially on coarse-textured or shallow soils. Inadequate rainfall limits plant growth and crop yields nearly every year in the region's semiarid areas. At other locations in the region, economic crop production is not possible without irrigation.

While the need for irrigation in the North Central Region is not as great as in regions to the west and south, the irrigated area in the region constitutes a significant portion of the irrigated land in the United States (Table 1). While only two states in the region rank in the top 10 in total irrigated area and in sprinkler irrigated area, the number increases to four when considering land irrigated with center pivot systems. However, 80 percent of the land irrigated in the region is concentrated in Kansas and Nebraska.

The land under irrigation in the North Central Region is about 22 percent of the total irrigated area in the United States. The percentage increases to 33 percent when considering only sprinkler irrigation and 56 percent for center pivot irrigation systems. About 50 percent of the area irrigated in the region is accomplished with sprinkler systems and 77 percent of the sprinkler area is irrigated with center pivot systems (Table 2). The popularity of sprinklers in general and center pivots in particular has implications for the types of problems faced by irrigators in the region.

Energy consumption by these systems is generally greater than that by surface systems; therefore, energy management should be of utmost importance. Because sprinkler and center pivot systems provide the capability for improved water management, advanced water management concepts should be more easily applied in the region than in other areas of the United States.

Much of the existing information on irrigation water management has been developed by universities within the North Central Region and the U.S. Department of Agriculture. Additional data have been obtained in other areas of the country, much of which can be used to improve irrigation water management in the region.

Concepts in Irrigation Water Management

Irrigation water is applied in areas where natural precipitation and stored soil water are insufficient to meet crop water requirements during the growing season. This water is applied to the soil surface through various systems, some elementary and others more sophisticated. The

Table 1. Irrigated area, by state, in the North Central Region, 1981 (3).

State	Irrigated Area (1,000 ha)	Rank	Sprinkler Irrigated Area (1,000 ha)	Rank	Center Pivot Irrigated Area (1,000 ha)	Rank
Illinois	61.1	32	61.1	26	44.5	19
Indiana	44.5	34	42.1	28	31.1	21
Iowa	101.1	26	92.3	18	68.4	13
Kansas	1,254.2	6	461.8	7	426.9	2
Michigan	130.7	24	130.7	14	56.9	14
Minnesota	204.3	20	198.0	11	128.1	8
Missouri	175.7	23	75.7	19	51.4	17
Nebraska	3,201.2	3	1,371.3	1	1,095.5	1
North Dakota	80.4	27	62.1	24	56.3	15
Ohio	19.8	38	19.8	36	3.4	32
South Dakota	184.9	22	160.6	13	127.1	9
Wisconsin	102.4	25	102.4	17	53.8	16
Region's percent of U.S. total	22		33		56	

Table 2. Irrigated area in the North Central Region (3).

Region*	Irrigated Area (1,000 ha)	Sprinkler Irrigated Area (1,000 ha)	Sprinkler Irrigated (%)	Center Pivot Irrigated Area (1,000 ha)	Center Pivot Irrigated (% of Sprinkler)
Northern Plains	4,720.7	2,055.8	44	1,705.8	83
Lake States	437.4	431.1	99	238.8	55
Corn Belt	402.3	291.0	72	198.8	68
Total	5,560.4	2,777.9	50	2,143.4	77

*Northern Plains—North Dakota, South Dakota, Nebraska, Kansas; Lake States—Minnesota, Wisconsin, and Michigan; Corn Belt—Iowa, Missouri, Illinois, Indiana, and Ohio.

disposition of water during and after an irrigation is called the on-farm water balance.

Inefficient use of irrigation water results from the physical condition of the off-farm conveyance system and the on-farm irrigation system as well as the improper management of these systems. Efficient use of irrigation water may also be influenced by institutional and social factors.

Implicit in the on-farm water balance is the concept that some deep percolation is needed to maintain a favorable salt balance in the plant root zone. Excess salts must be leached out of the root zone if crop productivity is to be maintained. The fraction of deep percolation needed to displace salts below the root zone, called the leaching requirement, must be considered an integral part of the irrigation water requirement. Salinity control is particulary important in areas where annual rainfall is insufficient to achieve the required salt balance.

Water Use Efficiency

Valid comparisons of water management strategies as these strategies affect physical and economic productivity require quantitative parameters. Where water supplies are scarce or where water is the primary limiting input to agricultural production, a physical parameter is preferable because its value does not change with economic conditions.

Water use efficiency has been used for the past quarter century to evaluate the effectivenss of alternative management practices and the efficiency of new cultivars. Water use efficiency is defined as the marketable crop produced per unit of water consumed in evapotranspiration. It can be used to compare production per unit of water consumed under dryland, limited irrigation, or full irrigation.

For meaningful comparisons, water use efficiency must also be based on water consumed during specific periods in the crop production cycle, such as planting to harvest or harvest to harvest:

$$\text{WUEnet} = \frac{\text{Crop production in kilograms}}{\text{Evapotranspiration, planting to harvest, in cubic meters}} \quad [1]$$

$$\text{WUEgross} = \frac{\text{Crop production in kilograms}}{\text{Evapotranspiration, harvest to harvest, in cubic meters}} \quad [2]$$

A similar term can be defined for increased crop production from irrigation (*1*):

$$\text{WUEirrig} = \frac{\text{Yield (irrigated)} - \text{Yield (dryland), in kilograms}}{\text{ET (irrigated)} - \text{ET (dryland), in cubic meters}} \quad [3]$$

Net water use efficiency is the most common form of reporting observed data. Gross water use efficiency is most often reported under dryland farming. Gross water use efficiency evaluations are needed under irrigation because evaporation following preplant irrigations can represent a significant portion of evapotranspiration from harvest to harvest.

As defined, water use efficiency will be zero if there is no marketable crop yield. Figure 1 shows examples of water use efficiency for grain sorghum and corn to illustrate typical relationships between water use efficiency and crop evapotranspiration. Under dryland farming, the net water use efficiency for both grain sorghum and corn is low when both yield and evapotranspiration are relatively small. Limited irrigation can increase net water use efficiency; however, the highest water use efficiency is always obtained at the highest yields. When adequate water is available, the water use efficiency for corn often exceeds that for grain sorghum. But under water-limiting conditions, when evapotranspiration is lower, the water use efficiency for grain sorghum may exceed that of corn.

Irrigation Efficiency

Significant improvements in understanding the interaction between on-farm irrigation practices and basin hydrology have occurred in recent years (*9*). Today it is recognized that irrigation system losses on the farm and in the delivery system are not necessarily losses to the basin. In

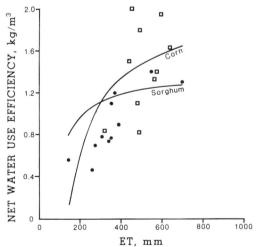

Figure 1. Net water use efficiency for corn and grain sorghum (*11*).

general, a portion of water that remains on or below the land surface after it is used for irrigation is recoverable for reuse within the basin. This fact imposes an important distinction among the various forms of water disposition on an irrigated farm. Water transferred to the atmosphere as a result of evaporation and transpiration and water evaporated during irrigation are irrecoverable. Runoff and deep percolation may be recovered with additional energy inputs unless the water is displaced to saline bodies where recovery is not economically feasible. Thus, a portion of the water diverted at the farm and not used in the evaporation and transpiration processes, while being lost to the farm, returns to the basin and becomes part of the water supply of another user. Examples of "lost" water include phreatophyte consumption, evaporation, and irrecoverable groundwater.

Quantitative parameters also are needed to evaluate the impacts of irrigation water management systems and practices. Several efficiency terms are commonly used by irrigation technologists in designing and evaluating irrigation systems and projects (2, 12).

Part of the irrigation water diverted is not consumed and usually returns to the original or another usable source. This component is frequently ignored by many people discussing water issues, which often creates confusion. "Effective irrigation efficiency" was defined several years ago to account for irrigation return flow (9):

$$E_{ie} = E_i + r(1 - E_i) \tag{4}$$

where E_{ie} is the effective irrigation efficiency; E_i is irrigation efficiency, which is the ratio of irrigation water consumed by crops on a farm or project to the water diverted from a river or natural source into the farm or project canals; and r is the fraction of excess water that returns to the river or groundwater for reuse (9). Basically, this term is the ratio of crop evapotranspiration plus return flow to the volume of water diverted. A more realistic measure of the efficiency with which water is used in irrigated agriculture is the ratio of crop evapotranspiration to net depletion of water rather than water diverted.

Net farm or project irrigation efficiency is the ratio of irrigation water consumed by crops on a farm or project to the net depletion of usable water in a river basin or groundwater system. Net irrigation efficiency (E_{net}) can be calculated as follows:

$$E_{net} = \frac{V_{et}}{V_{dep}} = \frac{V_{et}}{V_{et} + (1 - r)(V_d - V_{et})} \tag{5}$$

where V_{et}, V_{dep}, and V_d are the volumes of water consumed in evapotranspiration, the volume depleted, and the volume diverted, respectively.

This efficiency term is identical to E_i if there is no return flow to a reusable source (i.e., if $r = 0$). The magnitude of E_{net} is similar, but not the same as E_{ie}. E_{net} is an improvement in reporting irrigation effectiveness over previous terms, and its use should result in fewer misunderstandings of irrigation systems and practices.

Technological Developments in Irrigation Management

Irrigation Water Management

The concept of irrigation scheduling has received much attention in recent years. Irrigation scheduling is defined as predicting the time and amount of the next one or more irrigations, taking into account expected precipitation (10). The most common management technique, where water is not limited and its cost is either low or not based on volume, is to eliminate water as a production-limiting variable. The negative effects of applying excess water (deep percolation losses) are reduced by delaying irrigations until the soil water depletion is sufficient to permit storage of the next application. Plant water stress is avoided by irrigating before crop yield and/or crop quality are reduced because of inadequate soil water. Irrigation scheduling considers rainfall, evapotranspiration since the last irrigation, allowable soil water depletion at the present growth stage, and expected rainfall before the next irrigation. Irrigation scheduling requires that irrigators make decisions daily or weekly. Use of irrigation scheduling thus will, in most cases, reduce deep percolation, thereby improving irrigation efficiency, which lowers irrigation requirements.

A recent survey in California revealed some of the problems and challenges with the adoption of new irrigation scheduling technologies (4). Conclusions reached in that study included:

1. Irrigation scheduling programs that supply only current evapotranspiration data on a regional basis without providing specific on-farm services are not effective unless they are accompanied by intensive educational efforts.

2. An irrigation scheduling service is adopted when economic benefits can readily be identified by the grower. Thus, there is no question that inexpensive and readily available water reduced the incentive to adopt such a service. Potential scarcity alone stimulates significant interest. However, inexpensive, abundant water supply does not preclude interest in an irrigation scheduling service. Some growers with inexpensive, abundant water benefitted from an irrigation scheduling service by identifying and better managing problem soils, better organizing farm operations, and ultimately saving money.

3. An irrigation scheduling service can only increase growers' net income to the extent that use of some production inputs (water, energy,

fertilizer, etc.) decline or yields increase. There is a general skepticism among farmers that yields can be improved by altering irrigation practices. On the contrary, among those providing irrigation scheduling services, there is an overemphasis on the degree to which yields can be improved or water use reduced.

The role of the private sector in on-farm operations, such as irrigation system design, irrigation water management, timing, and evaluation, cannot be overemphasized. Private as well as agency scheduling service groups must be concerned about the probable acceptance of any new or improved technology. Survival of private service groups in particular depends upon usable, cost-effective technology that provides information that farmers want and need to manage irrigations (10).

Irrigation scheduling not only saves water but may also reduce peak electrical demand. The cost of electric power is determined by total energy use (kilowatt-hours) and by the maximum rate of power consumption (kilowatts) or peak demand. Irrigation electrical loads produce a high summertime power demand in heavily irrigated areas. The electric power suppliers must provide generating facilities and transmission lines capable of meeting the peak demand. However, many of these generating facilities remain unused during the eight or nine months when the pumps are not operating. Some retail power suppliers must pay a penalty to the wholesale supplier when the monthly peak load during the six winter months falls below a specified amount of the summer peak. The greater the difference between summer and winter loads, the greater the penalty.

A number of alternatives have been proposed and used to minimize peak demands (7, 18). The alternatives include (1) voluntary reduction in pumping during periods of high demand, (2) controlled interruption of pumping units, (3) water management integrated with load management, (4) redesign of systems for reduced pressure and discharge, and (5) an increase in pumping plant efficiency. Each alternative will reduce peak demand differently, and each will have different implementation costs.

In addition to the water savings resulting from the use of irrigation scheduling, the pollutants present in irrigation return flow will be reduced. In particular, the movement of nitrate nitrogen to groundwater may be reduced through better irrigation management procedures (14).

Limited Irrigation

Normally, when dryland farmers start irrigating, and if adequate water is available, they plant, fertilize, and irrigate for near-maximum yields. But recently, particularly in the southern Great Plains, reduced well yields have forced many farmers to practice limited irrigation. Escalating energy costs also have caused more concern about the amount of water pumped from wells with yields that have not decreased.

When irrigation water is applied at rates less than the crop evapotranspiration, two different situations may develop. If soil water is already depleted or is not sufficient to compensate for the difference between applied water and evapotranspiration demand, plant control of transpiration enters into play and water deficits develop. This situation, called deficit irrigation, should not be confused with the second possibility, where maximum evapotranspiration rates are met by a combination of insufficient irrigation water and soil water depletion. In the latter situation, the term "deficit" does not apply because no evapotranspiration deficit results under such irrigation regimes. In fact, in deep soils in semiarid regions, insufficient irrigation can and is practiced successfully to reduce peak irrigation requirements by conjunctively using irrigation and stored soil water. If the stored soil water is insufficient to meet evapotranspiration demand, plant water deficits develop under deficit irrigation even when practiced at a very high frequency (5). Through precise irrigation scheduling, limited irrigation could be used, resulting in some yield reduction. The magnitude of the yield reduction will depend upon the timing and degree of stress as well as the irrigation management procedure selected (6).

Farmers' primary objective in irrigated agriculture, as in other agricultural systems, is to maximize net returns. The economics of applying less water than required for maximum crop production thus must be considered. Numerous studies on maximizing crop production per unit of water used have been conducted over the last three decades. Most have focused on the relation (called water production functions) between applied water and/or evapotranspiration and yield as water supply declines below that required for maximum yields. Two recent reviews on water production functions have been published (8, 19). In general, the evapotranspiration-marketable yield relation for many crops is linear (while the irrigation amount-yield relation is typically nonlinear because of irrigation losses). Thus, reductions in evapotranspiration below the maximum do not increase water use efficiency. There may be some crops where reductions in evapotranspiration do not reduce yields in the same proportion as might be inferred from a summary of Israeli investigations on yield versus applied water in 16 crops (15).

In areas where the irrigated land area is fixed, high application efficiencies are attainable, and sufficient water is available, profit usually increases until yield approaches the maximum level (16). However, as the costs to pump and distribute water increase, the optimum yield may be below the maximum yield.

For a limited water supply, relative to available land, the economic objective is to maximize profit over the available area. When application efficiencies are high (greater than 70 percent), the optimum seasonal irrigation depth again often approaches the amount needed to achieve maximum yields (*16*). For systems with lower efficiencies, the problem becomes more complex because yield versus applied water functions are difficult to define. With the economic optimum frequently near the maximum yield, the water management criteria of primary interest tend to be those most likely to efficiently achieve yields close to the maximum (*17*).

There are limitations on implementing water production functions for generalized use, however. Because of the empirical nature of such functions, their transferability may be questionable. A California study questioned whether water production functions could be transferred in view of their empirical nature and concluded that knowledge of water production functions is not adequate to provide a basis for public policies that promote deficit irrigation (*19*). Additional limitations identified in the study were as follows:

1. Programming moisture stress entails substantial uncertainties unless actual evapotranspiration and the predicted evapotranspiration deficit can be monitored closely and precisely.

2. The relation between applied water (the variable under the control of the grower) and evapotranspiration is affected by so many site-specific variables that it would require designing deficit irrigation programs for each individual farmer and perhaps for each field.

3. The implications of risk for the water production functions have been explored in a preliminary way only. In general, the risks associated with deficit irrigation are much more significant than those associated with applying excess water.

The degree of risk varies with the cropping system, and the level of risk tolerable to the farmer will vary with economic factors. However, with the intensive cropping systems characteristic of irrigated agriculture in the North Central Region, providing adequate water for maximum evapotranspiration is perhaps the best interim policy until a different strategy is developed, based on specific, carefully researched information.

Optimal irrigation scheduling thus requires a high level of water management to ensure against overapplication under conditions of adequate water supply, yet careful control to minimize yield reductions resulting from water stress when the supply is limited. In either case, careful water management is required.

Improvements in Water Storage and Conveyance Systems

Most water used for irrigation in the North Central Region is from groundwater sources. There are relatively few surface water delivery projects in the region, and those are in the Northern Plains, predominately Kansas and Nebraska. Groundwater aquifers are one of the most efficient ways of storing water from a water balance viewpoint, but they may not be the most efficient from an energy conservation viewpoint.

Greater conjunctive use of surface water and groundwater storage systems will be needed in the future to more fully develop and manage limited water resources.

Improving water conveyance and on-farm water distribution systems can greatly increase our ability to irrigate efficiently and avoid plant water stress. Improvements can assure maximum delivery of water to users, reduce or eliminate consumptive waste along canals and on farms, and reduce energy costs for pumping. Improving farm irrigation systems to provide better and more timely water control can result in direct benefits to individual water users in terms of reduced energy costs, higher or better quality yields as a result of improved water management, and less labor. However, these improvements usually require large capital investments and generally increase overall irrigation costs. Improvements in water conveyance and on-farm distribution usually increase the proportion of water diverted or pumped that is made available to primary users in a project. But if these increases are used to irrigate additional land, they may reduce by a like amount the return flow from seepage and spillage.

Improvements in Irrigation Systems

Surface irrigation systems, primarily furrow systems, now irrigate about half of the irrigated land in the region. Approximately 94 percent of this area is in Nebraska and Kansas. Furrow irrigation on sloping fields permits uniform water applications, but unless the surface runoff can be recirculated or reused, the attainable application efficiency probably will not exceed 65 percent. Reuse systems are common in Kansas and Nebraska.

Efficient irrigation is currently easier to achieve with sprinkler systems. With modern automated controls, sprinkler systems provide good control of water applications. The exception perhaps is under windy conditions. Currently, many surface irrigation systems are being converted to sprinkler systems because of the lower labor requirements and ease of automation.

The center pivot sprinkler system has become so common it needs little description. Center pivot systems are now the most common sprinkler systems in the United States. Development of dependable automatic equipment and electronic controls has resulted in a system that enables uniform application of small amounts of water. These systems also enable farmers to irrigate rolling or sandy lands that were previously unsuitable for irrigation. The irrigation industry also has become well organized to deliver, install, and maintain center pivot systems. In addition, the technology now includes corner systems, which can irrigate a portion of the previously unirrigated corners of a field, and the continuously moving lateral system, which can irrigate essentially all of the field. These systems can either use an open ditch feed, flexible hose connection to a buried pipe, or a system that automatically connects itself to a buried pipe.

The continuously moving systems (center pivot or lateral move) also can operate at a significantly lower water pressure. While these lower pressures may cause potential runoff problems because of higher application rates, modified tillage practices can be incorporated into the farming program to reduce runoff losses. For example, microbasins can be used to store excess water near the point of application until the infiltration is complete (13). The water application device can be placed below the crop canopy, thereby reducing evaporation loss and improving irrigation efficiency. In addition, the basins store all but the largest rains, thus minimizing irrigation requirements.

In areas that receive significant amounts of rainfall in storms having large depths, the basins may need to be reconstructed during the irrigation season. This can create problems under tall crops, such as corn. The pressure requirements of these ultra-low pressure systems is reduced to the friction losses in the system. Of course, elevation changes in the field will have to be accounted for. Pressure regulators have been or can be developed to ensure uniform application of water.

Trickle irrigation has expanded rapidly during the past decade and achieved a high degree of sophistication. It still is used on only about 1 percent of the nation's irrigated cropland, however. Trickle irrigation is expensive. Its use is mainly for high-value row and orchard crops and where water costs are high or supplies very limited. Studies have shown that evapotranspiration for row crops is about the same with trickle systems as with other systems. Exceptions include new orchards, citrus groves, and vineyards. Clogging remains the number one problem with trickle systems, and expensive filtration systems are needed in most cases to prevent clogging. While systems may be designed for 95 percent uniformity, many do not achieve this because of clogging, manufacturer's variability, or poor management.

Greater irrigation efficiency is not automatically achieved by installing an improved system because irrigation efficiency depends upon the potential efficiency of the system and the way the system is managed and operated. Also, as indicated, an increase in irrigation efficiency may not proportionally increase the net supply of water available to agriculture.

Improved irrigation systems require large capital investments and generally increase annual irrigation costs. Because of better water control and indirect benefits, however, farmers will continue to improve their irrigation systems. Major changes will occur first in water-short areas, or where water supplies are limited, and where high-value crops can be grown.

Energy Management

Modern irrigation equipment and technology has the capability of applying water precisely. Yet this equipment and technology requires a significant expenditure of energy for its use. Because of the energy requirements, the profitability of irrigated agriculture using pumped water is highly sensitive to energy prices as well as farm commodity prices. Efficient use of energy in irrigation is important in today's economy and will become even more so in the future.

Many energy saving practices are available to the irrigator. Most of these practices require capital investments to improve the system's operation. From the standpoint of action to reduce energy use in irrigation, therefore, it is not sufficient to consider only the energy savings from particular adjustments or changes. One must also consider the economic feasibility of such alternatives. Several researchers are working on these questions in the North Central Region.

Water Quality

Irrigated crop production requires many of the same management practices as dryland agriculture. However, some practices, particularly the application of nutrients and pesticides, are likely to be more intensive in areas where crops are irrigated. While this generally results in higher crop yields, it also increases the possibility for negative impacts on the environment. For example, excessive use or mismanagement of water, nutrients, or pesticides or improperly designed or managed irrigation systems can create nonpoint source pollution problems.

One of the major problems facing modern agriculture is that of developing management practices that maintain high production while minimizing environmental hazards. Such practices must fit the varying conditions imposed by soil, climate, and crop. An optimum set of practices at one location may be disastrous at another. Appropriate management practices for a given set of conditions can be defined and implemented much better when the people who manage agricultural production and those responsible for environmental protection understand the general interrelationships between soil, plants, and water and irrigation management practices and limitations. A grasp of a few basic concepts is helpful in understanding how irrigation practices affect the quality of surface runoff and subsurface drainage water from irrigated land. Improved understanding also enhances the ability to alter practices as a means of improving the environment and maintaining economic production.

Water quality problems resulting from irrigation return flows and management alternatives for the control of these problems have been identified in many areas, particularly the Great Plains (*14*). The primary pollutants in irrigation return flow include sediments and colloids, nitrates, phosphates, pesticides, salts, and organics. Other factors must also be considered when determining water quality impairment, however. Among them are (1) the complexity of pollutant loading processes, (2) the large number of localized conditions that affect pollutant loading, and (3) the contribution of the same kind of pollutants from various sources, both natural and man-induced. Natural processes supply some known pollutants. In some cases, the effects of these can be separated from man-induced pollutants through the use of site-specific information.

Runoff and deep percolation on irrigated cropland can rarely be eliminated. But it can be reduced, substantially in some cases, by carefully selected combinations of management practices. Several management options to reduce pollutants in irrigation return flow, including irrigation system management, on-farm water management, soil management, nutrient management, and pesticide management, have been evaluated (*14*). In most cases, the technologies that can improve water use on the farm can also reduce the pollutants in irrigation return flow.

REFERENCES

1. Bos, M. G. 1980. *Irrigation efficiencies at crop production level.* International Commission on Irrigation and Drainage Bulletin 29(2): 18-25, 60.
2. Bos, M. G., and J. Nugteren. 1978. *On irrigation efficiencies.* International Institute for Land Reclamation and Improvement, Wageningen, The Netherlands. 142 pp.
3. Brantwood Publications. 1981. *Irrigation survey.* Irrigation Journal 31(6): 59-66.
4. Fereres, E. 1981. *Preliminary evaluation of irrigation scheduling services and farmer practices in selected areas of California.* Final report to the California Department of Water Resources, Agreement No. B54104. Department of Land, Air and Water Resources, University of California, Davis. 52 pp.
5. Fereres, E., et al. 1978. *A closer look at deficit high frequency irrigation.* California Agriculture 32(8): 4-5.
6. Gilley, J. R., D. L. Martin, and W. E. Splinter. 1980. *Application of a simulation model of corn growth to irrigation management decisions.* In D. Yaron and C. Tapiero [editors] *Operations Research in Agriculture and Water Resources.* North-Holland Publishing Co., New York, New York. pp. 485-500.
7. Heermann, D. F., and H. R. Duke. 1978. *Electrical load management and water management.* In Proceedings, Irrigation Association Technical Conference. Irrigation Association, Silver Spring, Maryland. pp. 60-67.
8. Hexem, R. W., and E. O. Heady. 1978. *Water production functions for irrigated agriculture.* Iowa State University Press,, Ames. 215 pp.
9. Jensen, M. E. 1977. *Water conservation and irrigation systems.* In Proceedings, Climate-Technology Seminar. University of Missouri, Columbia. pp. 208-250.
10. Jensen, M. E. 1981. *Summary and challenges.* In *Irrigation Scheduling for Water and Energy Conservation in the 80's.* American Society of Agricultural Engineers, St. Joseph, Michigan. pp. 225-231.
11. Jensen, M. E. 1983. *Water resource technology and management.* In Proceedings,

RCA Symposium: Future Agricultural Technology and Resource Conservation. Iowa State University Press, Ames, Iowa. In press.

12. Jensen, M. E., editor. 1974. *Consumptive use of water and irrigation water requirements.* American Society of Civil Engineers, New York, New York. 227 pp.

13. Lyle, W. M., and J. R. Bordovsky. 1979. *Traveling low energy precision irrigator.* In Proceedings, Irrigation and Drainage Specialty Conference. American Society of Civil Engineers, New York, New York. pp. 121-131.

14. Quinn, M. L., editor. 1982. *Strategies for reducing pollutants from irrigated lands in the Great Plains.* Water Resources Research Center, University of Nebraska, Lincoln. 188 pp.

15. Shalhevet, J., et al. 1981. *Irrigation of field and orchard crops under semi-arid conditions.* International Information Center, Bet Dagan, Israel. 132 pp.

16. Skogerboe, G. V., et al. 1979. *Potential effects of irrigation practices on crop yields in the Grand Valley.* EPA-600/2-79-149. U.S. Environmental Protection Agency, Washington, D.C. 194 pp.

17. Stegman, E. C., et al. 1981. *Irrigation water management—adequate or limited water.* In Proceedings, Second National Irrigation Symposium. American Society of Agricultural Engineers, St. Joseph, Michigan. pp. 154-165.

18. Stetson, L. E., et al. 1975. *Irrigation system management for reducing peak electrical demands.* Transactions, American Society of Agricultural Engineers 18(2): 303-306.

19. Vaux, H. J., et al. 1981. *Optimization of water use with respect to crop production.* Final Report to the California Department of Water Resources, Agreement No. B53395. Department of Land, Air, and Water Resources, University of California, Davis. 174 pp.

3

Water Quality: A Multidisciplinary Perspective

Lee A. Christensen

Interactions between agricultural production and environmental quality in the United States have been studied intensively, particularly in the past 10 to 15 years. Much of this research coincided with ambitious federal regulations to curb pollution from agricultural sources.

What follows is an overview of selected actions during the past 10 to 15 years that focused on the relationships between nonpoint source pollution from agriculture and water quality. Highlighted are approaches and lessons learned from chemical/biological and socioeconomic research.

Research and planning activities related to the technical aspects of water quality involve many organizations and perspectives. My discussion is organized around two themes, the nature of the activity and the responsible organizations. Attention centers on problem and process definition and planning guidance, implementation efforts, and evaluation and research efforts. A caveat is in order. The discussion is not intended to be all-inclusive, but focuses on representative activities.

The geographic focus is the eastern, humid region, of the United States. Water quality problems associated with nonpoint pollution from irrigation, timber production, and mining are beyond the scope of this paper.

Overview of a Decade of Water Quality Activities

The Emergence of the Issue

Early research by the American Association for the Advancement of Science described agriculture's relationship to environmental quality

36

(*12*). How agricultural wastes—livestock wastes, sediment, nutrients, and pesticides—damaged the environment were first documented in a general way (*115, 116*). Although the nature of agricultural pollutants was established, serious attention to the impacts of these pollutants on water quality was not forthcoming until 1972 with the impetus given passage of the Federal Water Pollution Control Act Amendments of 1972 (P.L. 92-500), particularly Section 208. These amendments established terminology and goals for the control of agricultural nonpoint source pollution.

The concern in the early 1970s was the control of point sources of pollution from industries and municipalities. However, there was recognition that the ambitious goals of "fishable and swimmable" waters by 1983 would never be met without reductions in nonpoint source pollution, to which agriculture was recognized as a major contributor, especially from animal wastes, sediment, and erosion (*15, 23, 25, 100*). It was also thought that in many cases there would likely be no noticeable water quality improvement from 1977 to 1983 because of the small amount of pollutants removed from regulated point sources compared with the nonpoint source loadings (*24*).

The initial responses of agencies charged with controlling agricultural nonpoint source pollution reflected their orientations from previous responsibilities. The Environmental Protection Agency (EPA) was oriented to mandatory rather than voluntary controls, a legacy from its point source control program. Some U.S. Department of Agriculture (USDA) agencies had years of experience with soil conservation technologies and presumed that the application of these techniques would generally be sufficient to meet water quality goals.

Those responsible for early efforts to control agricultural nonpoint source pollution quickly learned that little was known about relationships between agricultural practices in a particular field and the water quality of nearby streams. Related to this was the view that there was an inherent conflict between the goals of sustaining agricultural production and controlling agricultural nonpoint source pollution. Studies such as one undertaken by Cornell University for EPA (*45*) were intended to sharpen understanding of the relationships between soil conservation practices and water quality practices.

Since the mid-1970s, Section 208 planning, as well as state and federal programs, have attracted much attention and many dollars to solutions of agricultural nonpoint source pollution problems. Conferences, research meetings, and countless individual research efforts have been directed at identifying the nonpoint pollution problem and developing techniques for its control. Regulation of certain pesticide products produced environmental payoffs. But despite the time and money spent on agricultural nonpoint source pollution, recent U.S. Government Accounting Office (GAO) studies suggest that sediment and nutrient problems persist at levels comparable to 10 to 12 years ago (*109, 111*). As might be expected, given the complexity of the problem and the sudden growth in public concern, the main advances have been made in gaining better scientific understanding of the agricultural nonpoint source pollution processes, along with advances in modeling and measuring pollution movements and pollution control systems. Economic analysis of the costs and benefits of alternative control practices and policies has also improved.

Dimensions of Water Quality

What is meant by a water quality problem associated with agricultural nonpoint source pollution? Does it mean only the presence of undesirable materials in water resources, or is it a relative term that has to be related to impairments in use? To answer these questions, agricultural nonpoint source pollution and water quality need to be defined. Nonpoint source pollution has been defined as pollution arising from ill-defined and diffuse sources, such as runoff from cultivated fields, grazing lands, or urban areas (*94*). Agricultural nonpoint source pollution excludes runoff from urban areas, mining and construction activities, and highways, and for purposes of this paper, pollution from logging activities and streambank erosion. Nonpoint source discharges have been characterized as diffuse in nature, occurring primarily during rainfall events when runoff from the land surface carries sediment, sediment-absorbed chemicals, and dissolved chemicals into water systems. These discharges are also stochastic and dynamic in nature and have multimedia dimensions (*3*).

The most serious and most common types of pollutants found in agricultural nonpoint sources have been identified as bacteria, nutrients, dissolved solids (salinity), suspended solids (sediment), pathogenic organisms, and toxic materials. Runoff generally increases the level of bacteria, sediment, nutrients, and pesticides in receiving water, and irrigation return flow increases the level of dissolved solids, nutrients, and pesticides. Pesticides are generally considered the largest source of toxic pollutants in agriculture. Although concentrations and persistence of most pesticides now used are low, some chemicals are so highly toxic that even relatively low levels are cause for concern because of their impacts on fish and other aquatic forms. Leaching of plant nutrients and runoff can pollute groundwater with nitrates and cause accelerated surface water eutrophication. Dissolved solids (salinity) can corrode metals and adversely affect drinking water, livestock, and irrigated crops. Sediment can damage fish spawning areas, clog river channels, fill lakes, impair recreational uses, and carry absorbed pesticides and nutrients (*100*).

Citizens and officials originally faced a monumental task in ranking and finding a common definition of the nonpoint source pollution problem. A number of studies identified technical parameters related to water quality, but lacked provisions for measuring water quality in a manner directly useful for evaluating alternative projects (56, 81). An alternative was to define water quality in terms of major uses, including irrigation, municipal and industrial uses, recreation, ecosystem maintenance, power generation, flood control, drinking, and aesthetics (43). The specification of water quality only in terms of technical parameters is compatible with a regulatory approach that does not recognize the variability of both the costs and benefits of uniformly meeting fixed standards. Defining water quality in terms of impaired uses recognizes the variability in nonpoint source pollution problems and the differences in costs and benefits associated with alternative solutions.

What are Best Management Practices?

The focus of much effort in the control of agricultural nonpoint source pollution is the identification and evaluation of best management practices (BMPs). This term, first introduced in Section 208 of Public Law 92-500, represents an engineering concept that is now embodied in the lexicon of agricultural nonpoint source pollution control. BMPs have been defined as: "A practice or combination of practices that are determined by a state or designated areawide planning agency to be the most effective and practicable (including technological, economic, and institutional considerations) means of controlling point and nonpoint pollutants at levels compatible with environmental quality goals" (94). BMPs can be evaluated as individual practices. However, solutions to many water quality problems often require the concurrent use of more than one practice. Each farm, moreover, has a variety of water quality problems that require the use of several BMPs. Therefore, it is important to have the analytical tools for evaluating systems of BMPs as well as individual practices. Some criteria suggested for evaluating BMPs are agronomic effectiveness, environmental effectiveness, economic feasibility, social acceptability, and implementability (4).

Klaus Alt and associates (2) pointed out that there are difficulties in using the term "best." Nonpoint pollution abatement policies, including BMPs, they suggest, should be judged by their contribution to improving resource allocation in the economy. Conservation practices or policies thus can be judged "best" only if they improve resource allocation. Others (4) have noted difficulties in defining "best." Best for whom? For what purpose? At what scale?

BMPs focus on reducing pollutant loading at the source to meet certain goals without any explicit consideration of the impacts on impaired use of the streams. A concept of in-stream water quality management has been developed as an alternative strategy to source control (80). Source control strategy is based upon standards such as soil loss limits or best management practices. These standards are not necessarily related directly for water quality goals. In-stream water quality management is a strategy of determining water quality goals by examining pollution effects and other considerations in developing a resource management plan for achieving these goals.

One of the advantages of an in-stream water quality management plan is flexibility; approaches can be changed if water quality goals are not met. With source control, a plan is typically applied without analyzing changes in water quality. One of the deficiencies of the in-stream water quality management approach is that it requires much closer coordination between all parties involved in defining water uses and the management alternatives and goals. The higher costs of in-stream water quality management approach make it best suited to streams with valuable uses or important management problems.

Problem Definition and Planning Guidance

The passage of Public Law 92-500 and subsequent legislation and regulations gave new focus and impetus to academic, state, and federal agencies to develop techniques and approaches for reducing agricultural nonpoint source pollution.[1] Some efforts were shared by various agencies and universities; others were conducted independently. Much initial research was directed toward defining the problem and the processes involved, and analyzing the likely impacts of selected controls and policy options. Following are some representative examples of federal and state programs.

U.S. Environmental Protection Agency. EPA has played a key role and handled major responsibilities. The agency's involvement in research and planning for reducing agricultural nonpoint source pollution was required by Public Law 92-500, particularly Section 208. This law required each state to prepare water quality management plans, which needed the approval of EPA. EPA supported research in other agencies and at universities to develop information to aid states' efforts. The agency also developed a program at its own research laboratories to define agricultural nonpoint source pollution processes and to evaluate control techniques.

Several important modeling activities were developed at or for the Athens Environmental Research Laboratory in Georgia. The laboratory also initiated a multidisciplinary field study in the Four Mile Creek

---

[1]Concurrent with domestic U.S. effort was the 1972 U.S.-Canada Great Lakes Water Quality Agreement that established the Pollution from Land Use Activities Reference Group (PLUARG). As part of these activities, pilot watershed studies were established in Canada to quantify the Canadian agricultural pollution loadings to the Great Lakes (*20*).

watershed in Iowa, designed to help EPA better understand the relationships between practices on the ground, water quality, and impacts on farmers (5, 6, 69). Research at the Robert S. Kerr Laboratory, Ada, Oklahoma, included control of nonpoint source pollution from livestock production and irrigated agriculture. Work at the Corvallis, Oregon, laboratory included studies of ecological effects (77). EPA regional offices participated in such activities as the Indiana Black Creek project, an early, detailed study of agriculture's contribution to water quality degradation and reduced environmental quality (59). The study involved the Chicago regional office of EPA, the Soil Conservation Service, the Allen County (Indiana) Conservation District, Purdue University, and the University of Illinois.

U.S. Department of Agriculture, Agricultural Research Service (ARS). Prior to the 1970s, ARS had conducted extensive research on soil erosion and chemical transport related to agricultural production. With enactment of Public Law 92-500, ARS redirected some of its expertise to water quality problems. In the mid-1970s, ARS, in cooperation with EPA, prepared guides for planners to provide assistance to individuals involved in preparing 208 plans (103, 104). These guides focused on both nonpoint source pollution from crop production as well as from livestock production (106). They addressed primarily the physical science aspects. (Brief sections on economics prepared by USDA's Economics Research Service were also included.) ARS was also involved in cooperative research with EPA's Athens Research Laboratory on several watershed studies in Georgia (91).

ARS combined the individual research efforts of many of its scientists into models for use in agricultural nonpoint source management at both field-scale and watershed-scale levels. The first model, termed CREAMS, is a field-scale model for determining chemicals, runoff, and erosion from agricultural management systems (55). CREAMS took existing physical base models, particularly those that could easily be modified and improved upon, and combined them into a model for estimating runoff as well as sediment, plant nutrient, and pesticide movement in a field. A small watershed area model (SWAM) is currently under development.

U.S. Department of Agriculture, Economic Research Service (ERS). ERS has been involved in economic research related to agricultural nonpoint source pollution since the mid-1970s, primarily in its Natural Resource Economics Division. Early efforts were concentrated on identifying technologies for controlling agricultural pollutants (108), evaluating the environmental impacts of alternative cropping practices in an Iowa river basin (1), and assessing the impacts on farm income of constraints on soil loss, fertilizer use, and land use (*54*). A literature review provided background for ongoing research on the socioeconomic impacts of practices to reduce agricultural nonpoint source pollution (*76*). The potential of soil taxonomy as a technique for identifying soils suitable for minimum tillage technology was investigated (*21, 22*). Research efforts also modeled linkages between land use practices and stream quality to define the parameters of water quality (*43*) and identified benefits associated with various practices (*44*). Studies have estimated the off-site benefits to recreation and property values from improved water quality (*10, 11*). Others have examined the attitudes of farmers toward various best management practices (*17, 19, 50*). ERS research for EPA examined the economic factors involved in animal waste management for water quality improvement (*18*). ERS and EPA organized a water quality monitoring and modeling workshop to bring model builders and users together to discuss alternative approaches (*30*).

ERS and ARS scientists have worked together on several research projects to develop techniques and evaluation procedures. One resulted in refined procedures for evaluating measures for the control of dissolved phosphorus. Whereas earlier linear programming models (*16*) assumed that all soils in a watershed had the same phosphorus enrichment ratio, later research (*74*) made use of a buffer curve technique that identifies dissolved and sediment-absorbed phosphorus losses specific to each soil/crop management combination. In conjunction with activities of the Soil and Water Resources Conservation Act (RCA), an Erosion Productivity Impact Calculator (EPIC) was developed cooperatively to help answer questions related to the long-term impact of soil erosion on soil productivity (*37*). This model, which is still being refined and revised, has six components, two of which are related directly to water quality issues: (1) tillage and residue management and (2) soil erosion simulation.

ERS has been involved in implementation and evaluation efforts, such as the Model Implementation Program (MIP) and the Rural Clean Water Program (RCWP). The agency conducts economic analyses of projects in areas selected for comprehensive evaluation. Other important ERS efforts involve cooperative activities in the river basin planning studies throughout the country, many of which have water quality components.

University Research. The focus of much university research on water quality issues has been on physical processes and on providing planning guidance, primarily through the development and application of data bases and chemical, biological, and socioeconomic models and evaluation procedures. Much of this research has been supported by federal (EPA, Hatch Act, National Science Foundation, U.S. Department of the Interior) and state funds.

Some examples of university research dealing with economic relationships between agriculture and water quality include that completed at the Center for Agricultural and Rural Development (CARD), Iowa State University, and the Department of Agricultural Economics, University of Illinois. These universities have completed numerous studies assessing the economic impact of soil erosion controls on a national or regional basis. Demand for agricultural commodities has been linked to sediment generated on cropland (*112, 113*). The economic impacts of erosion control on soybean and corn production in the Corn Belt have been analyzed (*97*). It was concluded that soil loss taxes were more effective than terracing subsidies in reducing soil loss. Six alternative approaches for controlling water pollution from agricultural nonpoint source pollution in the Corn Belt were analyzed at the University of Illinois (*88*). The policy options examined were an educational program, a tax credit for erosion control practices, a 50 percent subsidy for terracing, a requirement for a conservation plan to be developed, a requirement that plans be implemented, and a requirement that greenbelts be developed where needed. Both the economic impact and equity of these alternatives were examined. As noted previously, in-stream water quality management is a promising alternative to source control (*80*).

Cornell University conducted a large research project jointly with EPA to evaluate the effectiveness of soil and water conservation practices for pollution control (*47*). The goal was to identify the potential water quality effects of soil and water conservation practices and to describe the economic implications of their use. Moreover, researchers developed and tested methodologies that could be used to estimate the effects of practices on pollution from cropland.

Several studies have developed data bases for addressing water quality issues. These same studies have also tested evaluation procedures. One example is the Four Mile Creek watershed study in Iowa. Federal and state scientists and policymakers cooperated in this field evaluation project to improve the evaluation of models developed by EPA and others. Their purpose was to test the effectiveness of BMPs and to select BMPs for a particular watershed (*5, 6, 53, 69, 71*). Another such study (*93*) collected data from the Altamaha River Basin in Georgia for use in analyzing agricultural nonpoint source pollution issues in the southern Piedmont (*93*).

Water Quality Implementation and Evaluation Activities

Two major programs that have focused USDA and EPA efforts on the implementation and evaluation of agricultural nonpoint source pollution controls are the Model Implementation Program (MIP) and the Rural Clean Water Program (RCWP). Additionally, efforts initiated under

RCA have developed information useful in the planning of nonpoint source pollution control activities.

The Model Implementation Program

USDA and EPA began a joint effort in September 1977 to develop a three-year test called the Model Implementation Program (MIP). The primary purpose of MIP was to demonstrate the effectiveness, in small geographic areas, of concentrating and coordinating various USDA and EPA water quality management programs and to illustrate how the water quality management plans developed under Section 208 of the Federal Water Pollution Control Act Amendments of 1972 could be translated into action. Seven project areas were selected from a group of 50 applicants. MIPs were located in Indiana, Nebraska, New York, Oklahoma, South Carolina, South Dakota, and Washington. The projects were to evaluate (1) whether USDA and EPA agencies at the national, regional, state, and local levels could coordinate and accelerate their programs and activities among themselves and with local government and gain sufficient support and participation from local governments and farmers to install BMPs on the land; (2) whether sufficient BMPs could be installed under the MIP effort to reduce agricultural and silvicultural nonpoint source pollution and if these practices were capable of reducing nonpoint source pollution; and (3) what impact these BMPs had on water quality improvement (*60, 61*).

Because MIPs were not established as controlled experiments, they could not provide definite answers to questions about the cause and effect relationships between practices installed on the land and changes in water quality. However, the efforts helped develop closer working relationships between USDA and EPA and provided useful insights for RCWP (*51*).

The Rural Clean Water Program

RCWP contains both implementation and evaluation components. The program was authorized by Section 35 of the Clean Water Act of 1977 (Public Law 95-217), which amended Section 208 of Public Law 92-500. Although the act called for an extensive program, planned funding of $400 to $600 million per year never materialized. Finally, Congress appropriated $50 million to USDA in fiscal year 1980 to initiate RCWP on an experimental basis. Congress added another $20 million in fiscal year 1981.

The program is administered at the national, state, and county levels by the Agricultural Stabilization and Conservation Service (ASCS) and other USDA agencies. The Soil Conservation Service (SCS) coordinates technical assistance at both the project and national levels (*105*). The goals of the program are (1) to improve impaired water use and quality in approved project areas in the most cost-effective manner possible; (2) to assist farmers in reducing agricultural nonpoint source pollutants and in improving water quality in rural areas; and (3) to develop and test programs, policies, and procedures.

USDA has implemented the program in 21 selected project areas throughout the country. In each project USDA enters into 5- to 10-year contracts with owners and operators of lands identified as critical pollution source areas to install and maintain BMPs for reducing nonpoint source pollution. USDA also provides financial and technical assistance. Participation in the program is strictly voluntary. The maximum federal cost-sharing rate is 95 percent, and the maximum total expenditure is $50,000 per participant. These maximum cost-sharing rates and amounts are well above the ceilings for Agricultural Conservation Program (ACP) payments.

Three types of *ex post* evaluations of the experimental RCWP are in process: general, comprehensive, and program. General evaluations are designed to assess BMP application progress and to document related changes in water quality attributed to an RCWP project. Comprehensive evaluations are designed to assess the impact and effectiveness of representative projects and/or specific RCWP project components. Program evaluations are designed to determine the extent to which the RCWP is meeting its objectives, the effect of RCWP on its participants as opposed to the rest of society, and the relationship of costs to benefits (*30*).

Five of the 21 RCWP projects were selected for comprehensive evaluations (located in Idaho, Illinois, Vermont, South Dakota, and Pennsylvania); the remaining 16 will receive general evaluations. The overall program evaluation will be based upon the combined results of the comprehensive and general evaluations.

Comprehensive evaluations consider both physical and socioeconomic impacts and involve both modeling and stream monitoring activities. ERS is responsible for evaluating the socioeconomic impacts, which include (1) impacts on participants and local agriculture, (2) off-site and community benefits, (3) cost-effectiveness of practices, and (4) overall project benefits and costs. Noteworthy are the attempts to evaluate such off-site impacts as recreation benefits, residential property value improvements, and reduced water treatment and power generation costs.

The Soil and Water Resources Conservation Act

Public Law 95-192, RCA, directed the secretary of agriculture to develop a national soil and water conservation program to guide USDA's future conservation activities on the nation's private and other nonfederal lands and to update that program every five years. RCA activities have measured the nation's resource capacity and determined a research agenda for the 1980s and beyond for USDA, identifying both water quality and soil productivity as priority resource concerns.

RCA reports have appraised existing resource conditions, trends, projected resource problems, and needed activities (*100, 101, 102*). Major resource problems addressed were soil erosion, flood damages, water conservation, and water quality. Areas of the country with critical nonpoint pollution problems were ranked on the basis of social and environmental factors and the effects of pollutants on water quality.

Uses of Models and Modeling

Because of the diversity of factors in agricultural nonpoint source pollution in a particular geographic area or soil type, it is impossible to develop standardized sets of procedures or recommendations for controlling pollution. Because of the large costs of generating site-specific information, physical and economic models have been developed to explain the complex relationships between agricultural practices and the environment and to predict the impacts of alternative courses of action. Models, as abstractions of reality, permit simplification of real world phenomena so they can be handled more effectively. Models can be used to define cause and effect relationships, extend utility of limited and costly field data, and permit *ex ante* evaluation of nonpoint source pollution control practices and policies.

Three dimensions of the agricultural nonpoint source pollution problem have been noted: a physical/technical problem, an economic problem, and a social problem (*3*). Efforts to model the various dimensions of agricultural nonpoint source pollution have been summarized (*3, 45, 46*). Predictive equations for soil loss, water yields in the fields, and water and sediment routes into watersheds are models of individual pollutant transport and transformation processes that need to be linked in order to simulate overall systems behavior (*3*). G. W. Bailey and Robert Swank suggest the following characteristics for nonpoint sources assessment models: water balance descriptions, soil loss descriptions as a function of particle size, incorporation of pollutant attenuation and transformation processes in soil and water, multimedia pollutant transport capability, continuous simulation, cost-effective operation, algorithms sensitive to management options, universal language, easily obtained input data requirements, and the ability to operate at various scalar levels.

Models have been grouped for analysis of agricultural nonpoint source pollution into chemical transport models and planning and management models (*45*). Chemical transport models are means to predict the chemical loadings to water bodies resulting from alternative agricultural practices, thus providing a way to estimate the water quality implications of policy decisions. Planning and management models offer a way to look at the trade-offs between agricultural production and environmental quality and the related economic impacts.

Physical/Chemical Modeling

Much recent progress has been made in predicting the transport and fate of agricultural pollutants. An array of nonpoint source loading models and assessment and ranking methodologies have been developed (3). Models describing the chemical transport process and nonpoint source loading of streams are useful in problem definition and identification of BMPs. They range from screening-level sediment and associated pollutant loading models based upon the universal soil loss equation (USLE) (120), to continuous, processed-based, field-scale simulation models, such as the pesticide transport model (PTR) to estimate runoff, erosion, and pesticide losses from field areas (28); the agricultural runoff management model (ARM) to estimate runoff, erosion, and pesticide and plant nutrient losses from field areas (33, 35); and the chemical runoff and erosion from agricultural management systems model (CREAMS), a continuous simulation, field-scale model designed to estimate runoff, erosion, and absorbed and dissolved chemical transport (55). Other models include the generalized nonpoint source model (NPS) (34, 36), the Cornell nutrient simulator (CNS) (47), and the areal, nonpoint source watershed environment response simulation model (ANSWERS) (7), developed to estimate runoff, erosion, and sedimentation from basin-size areas. The agricultural chemical transport model (ACTMO) was developed to estimate losses from field- or basin-size areas (40). An event model has been developed to estimate pesticide losses from fields during single runoff-producing storms (13).

The process models have varying requirements for data and calibration. Several efforts have been identified (3) to provide multiple-year, high-quality runoff data bases (6, 38, 39, 91) for use in developing regional parameters for loading model hydrology, for testing predictive runoff quantity and quality model algorithms, and for developing new model algorithms and capabilities where deficiencies are revealed. Instream data are also being used to conduct limited tests of linked modeling systems (5, 52).

Economic Modeling

Many technical alternatives for improving water quality by controlling agricultural nonpoint source pollution exist that are sufficient to meet the various environmental requirements. One task for agricultural economists, however, is to identify which of the feasible technologies can be readily integrated into existing agricultural operations without (1) imposing severe financial hardships on individual farmers, (2) seriously disrupting regional or national agricultural production, or (3) creating serious inequities within the agricultural community or between farmers and the rest of society. Both physical and economic impacts should be measured in water quality modeling and evaluation. Measurements of economic impacts have been suggested, including (1) direct benefits and costs of both with and without changes in water quality and land and water use, (2) indirect benefits and costs, (3) administrative costs to implement and enforce water quality and conservation measures, and (4) distribution of benefits and costs (*64*).

A variety of economic techniques and models may be used to aid the selection and evaluation of BMPs. Analyses can range from partial budgeting done on a particular field, whole-farm budgeting or farm planning using budgeting or linear programming techniques, or complex regional and national models linked with physical/chemical models. Field or farm models are best suited for modeling farmer decision-making. They can also be used for assessing how farmers will react to various policy options or economic pressures and for financial analyses. Regional and national models are more useful for analyzing the aggregate production impacts and the distributional effects of alternative policies.

Farm-Level Models. Farm-level studies have been applied frequently to estimate the on-farm income impacts of individual or combinations of BMPs and thus provide guidance for practice selection. An Iowa study compared erosion control and reduced chemical use, concluding that soil loss restrictions had a much less severe impact on farm income than did reducing fertilizer use or banning pesticides (*63*). In a Wisconsin case study, the on-farm impacts of several levels of soil loss limits were estimated using a revenue-maximizing linear programming model coupled with a hydrology-based simulation model (*90*). The model predicted watershed sediment loadings associated with each land use configuration generated by the linear programming model. The study demonstrated the superiority of minimum tillage technology from both an economic and a sediment loading standpoint.

Studies to forecast which BMPs farmers will select often focus only on strict economic criteria. Several studies have indicated that farmers entertain objectives other than profit maximization that need to be taken into consideration in economic analyses (*17, 57, 71*). Farmers also have perceptions of the yield impacts and risks associated with new technologies that need to be considered in modeling (*68, 69*).

A New York dairy farm study (*83, 85*) used the universal soil loss equation and linear programming to determine the cost-effectiveness of alternative sediment control practices. Cost-effectiveness was determined by estimating reductions in soil erosion and sediment delivery if the practice or combination of practices were adopted and by estimating the cost of implementing the practices. The study distinguished between the control of soil erosion and the control of sediment. Soil erosion control is usually conducted on a field-by-field basis. Sediment control is not necessarily geared for field boundaries, but for soil movement from a group of fields to watercourses. Costs of erosion and sediment control were found to vary appreciably among farms. An absolute soil erosion limit could in some cases have severe effects on farm income. These findings are consistent with other studies (9, 52) in which linear programming models were used to estimate changes in farm income at various levels of erosion control. The other studies too demonstrated the variable impact of erosion controls on farm income.

A multidisciplinary study at Cornell University (47) provided planning guidance on a number of chemical/biological and economic issues. The study identified the potential water quality effects of soil and water conservation practices and the implications of their use as BMPs. Soil and water conservation practices were found to be most useful in controlling edge-of-field losses of sediment and total phosphorus from cropland. They will also reduce losses of moderately or strongly adsorbed pesticides, although there are more cost-effective controls. The practices are not so effective in reducing the movement of dissolved nitrogen carried in solution.

The costs and effectivenesses of selected soil and water conservation practices at the farm level are estimated in the economic section of the Cornell report (47). Rather than assuming that reductions in sediment would result in improved water quality as in several other studies (90, 117), the effectiveness of several practices was determined for some hypothetical field situations. The cost-effectiveness of practices for water quality improvement varied considerably among type of farm, climate, and physical properties of the field. Actual rankings of practice cost-effectiveness had to be determined by the individual situation. Based on a study of representative farms in New York, Iowa, and Texas, conservation tillage, no-till, contouring, and stripcropping were found to be prospective BMPs. The study also concluded that, in general, the higher the degree of effectiveness desired, the lower the cost-effectiveness.

In a planning guidance/methodological study (65), an analytical procedure was developed to examine simultaneously the water quality impacts of selected agricultural practices and the economic effects of these practices on the farmer. The procedure was used to evaluate individual soil and water conservation practices in the Black Creek watershed in Indiana. In this effort it was assumed that each practice evaluated would be implemented on the whole farm. The study did not compare actual farm plans where combinations of practices that vary from field to field would typically be used.

Watershed Models. Watershed models developed to estimate the impacts on agricultural nonpoint source controls and policies have typically been linear programming models. They are usually applied to solve a specific water quality problem through the control of agricultural nonpoint source pollution.

The effects on farm income of erosion controls have been studied on a watershed level (*1, 52*). Shifting row crops from erosive soils to soils with lower erosion potential is necessary to meet a variety of environmental constraints. Coupling appreciable reductions in gross soil erosion with only a modest reduction in gross watershed income is possible when it is feasible to shift crops among soil types.

Watershed studies completed in several states have compared the implementation and maintenance costs of soil and water conservation practices with the assumed benefits (*2, 86, 95*). Two studies have addressed practices to reduce sedimentation of downstream reservoirs (*2, 82*). A Texas study (*82*) showed that erosion control measures could not be justified economically even when social benefits and costs, including off-site effects, were considered. An Illinois watershed study (*78*) of nitrate leaching and sedimentation found that erosion reduction was economically feasible.

One issue commonly raised is the environmental trade-off between use of minimum tillage practices to reduce sediment losses and the possibility of increased pollution from pesticide residues. An optimal control strategy for sediment control may be suboptimal for pesticide control. A watershed study (*49*) concluded that erosion control strategies are compatible with pesticide exposure control at high levels of erosion control. Also, though pesticide use usually increases with no-till systems, the environmental characteristics of the pesticides are more important than the sheer volume of material.

Several watershed models have been developed to test and validate concepts and approaches. A method has been developed for assessing the water quality and socioeconomic impacts of agricultural practices (*119*). This approach traces agricultural nonpoint source pollution from its field origin through to a subjective evaluation of water quality impacts. It is feasible for state-level planning and for broad analysis of policy alternatives, including BMP identification. A procedure has been developed for evaluating the effectiveness and cost of alternative agricultural nonpoint source controls in watersheds where little water quality data exists (*84*). The procedure uses the concepts of pathway control, manageable nonpoint source load, cost-effectiveness, and cost equivalents to evaluate and rank different control practices or sets of practices.

Regional/National Models. Regional models are useful for a macro-scale evaluation of the likely impacts on aggregate economic measures of agricultural nonpoint source pollution policies. Indicators typically used are the level and distribution of agricultural income, the aggregate level of agricultural production and food prices, and regional shifts in agricultural production and income.

A national model developed at Iowa State University (*48*) was applied to evaluate the impacts of proposed environmental policies. The impacts of restrictions on cropland erosion and nitrogen fertilizer application were evaluated, and it was concluded that although limiting soil loss and nitrogen fertilizer applications had little national impact, regional changes can be severe. The study did not include in its analysis the transport of eroded soil or nitrogen to waterways.

A second study at Iowa State University (*114*) built upon the previous model by adding the transport of eroded soil to streams and reservoirs, a capability enabling the model to estimate directly water quality impacts. A general conclusion from that study was that uniform controls on erosion levels are expensive means of reducing river sediment loads because they are not limited to the most erosive cropland or that with high sediment delivery.

A regional linear programming model was developed to evaluate impacts in the Corn Belt of agricultural nonpoint source pollution control policies. The model included supply and demand relationships and measured the distribution of control costs among regions, farmers, and consumers. In its first application, the model evaluated the economic effects of alternative erosion controls, banning of insecticides and herbicides, and severe restrictions on rates of nitrogen application (*97*). These severe restrictions generally raised farm income, but also increased consumer costs. A subsequent application added erosion control policies, concluding that there would be small impacts on farm income, but that food prices could increase (*86*). A related study evaluated the impacts within the Corn Belt of the adoption of different soil erosion control policies by the different states (*79*). The study concluded that there would not be significant shifts in production among the states, that the most severe impacts would be at the substate level, and that farmers on the most erosive lands may be adversely affected.

Resources for the Future, Inc., developed a national network model that links both point and nonpoint sources of pollution at the county level to a national network of major rivers, lakes, reservoirs, and bays (*42*). The model was applied to evaluate agricultural sediment control policies in conjunction with point source controls (*41*). Results indicated that even after full implementation of point source and sediment control policies, about one-half of the nation's rivers will still experience violations of phosphorus and nitrogen standards.

What Was Learned? And What to Know?

The decade began with a goal to make America's streams fishable and swimmable. What has been accomplished? Considerable progress has been made in controlling pollution from point sources, but not pollution from nonpoint sources (*111*). Many research, planning, implementation, and evaluation activities have been conducted in recent years. Yet the problem persists, in part because in the last decade agriculture has become more intensive, resulting in increased nonpoint source pollution. Projections of future demands for agricultural production suggest that pressures on water quality will persist (*98, 99*).

What We Have Learned

The past 12 to 15 years may best be regarded as a learning period. University and agency research, planning, and implementation activities have provided and will continue to provide a much sounder scientific base for decision-making. Channels of communication between individuals and local, state, and federal groups concerned with water quality are much better defined as a result. There is greater realization that solving the problem requires a multidisciplinary approach. For example, merely knowing the physical/chemical processes by which nitrogen moves from the point of fertilizer application to a water body does not result in solutions for cleaning up a stream. Similarly, identifying a best management practice strictly on the basis of economic criteria may be inappropriate because of the noneconomic variables affecting a farmer's decision-making process. Great progress has been made in bringing together scientists from various disciplines to address a specific problem, such as in the Four Mile Creek study, the Black Creek study, the Model Implementation Programs and Rural Clean Water Program projects.

It is impossible here to summarize adequately all that has been learned in recent years about agricultural nonpoint source pollution and water quality. The following statements, however, represent my reflections on some of the lessons learned.

1. Agriculture still has significant adverse impacts on water quality. The contribution of agriculture to water quality problems, relative to point sources, is much better understood and appreciated than it once was. We are now much better equipped to define and model a problem and to select controls and evaluate their economic impacts. Data bases have been developed for model testing and evaluation. Yet the task of implementing a comprehensive program remains unfinished.

2. There is a better appreciation of the complexities of both the nature of the problem and the need for contributions from a variety of disciplines to identify and implement solutions. Uniform application of envi-

ronmental policies, such as a constant soil loss limit on all soils within a delineated area, is recognized as both inequitable and inefficient.

3. There is an increased realization that the solution to a problem in one area is not readily transferable to another. The variation in soils, climate, and other factors makes each problem area unique.

4. The distinctions between practices to control soil and water movement and other best management practices are much better defined than they used to be. Practices that keep soil on the field enhance productivity and also reduce the movement of sediment and sediment-transported chemicals (primarily phosphorus and pesticide residues) to streams. For other types of pollutants, soil conservation practices are not as effective, and alternative practices need to be evaluated.

5. There is not necessarily a direct relationship between the complexity and expense of control practices and effectiveness in reducing agricultural nonpoint source pollution. Simple adjustments in management practices can solve some problems better than can more costly structures.

6. There is a much better appreciation of the interaction and trade-offs between sources of pollution and their control. For example, control strategies designed for optimal erosion control may be less than adequate for controlling pollution from pesticides or nitrogen fertilizer.

7. There have been many applications of linear programming models to the analysis of BMP impacts at the farm, regional, and national levels. Sediment control has been modeled most frequently, followed by fertilizer and pesticide restrictions. The economic and equity impacts of policy options, such as taxes, are better understood. For example, soil loss taxes have been found more effective than terracing subsidies. Little has been done to incorporate risk and farmer attitudes into these models.

8. Economic forces contribute to improvement in water quality, perhaps as much as subsidies and threats of regulation. Studies have indicated (29) and experience substantiates that rising fuel costs have caused fairly rapid shifts to various forms of conservation tillage farming, which has brought reductions in erosion and sedimentation.

9. Successful solutions to water quality problems are not easy, quick, or inexpensive. They require a combination of good technical information, enough time to install practices and measure effectiveness, a monitoring program that can measure stream quality changes, sufficient money, and an overall evaluation program. Absence of any of these components can limit progress in water quality improvement.

What We Need to Learn

There are plenty of opportunities for future research. Needed is much better documentation of the water quality improvements resulting from agricultural nonpoint source pollution controls. Progress in water quality improvements has been reported (*23, 24, 25, 26, 27*), but my investigation has not found documentation of the aggregate results of applied practices. However, improvement in lakes and streams in certain areas has been observed, particularly by users of these resources. Much information has been developed to aid in selecting BMPs and measuring on-farm environmental and economic impacts. However, much more work is needed in estimating off-site costs and benefits of improved water quality due to control of agricultural nonpoint source pollution. This area has perhaps the greatest opportunities and challenges in the next several years. The empirical relationship between BMPs applied on a farm field and the quality of receiving streams needs to be better defined. Many questions remain about estimating sediment delivery ratios and other indicators of pollution transport from fields to receiving streams.

Looking to the Future

Finally, several emerging issues and factors will have impacts on water quality activities. There are uncertainties about the future direction of water quality laws and the level of commitment to water quality goals. EPA is about to issue proposed rule changes. The changes would replace uniform national water quality standards with differing state and stream reach standards, which represents a retreat from the goals of the Clean Water Act. There are questions about the future availability of funds for the 208 program. Currently, no funds exist for preparing 208 plans, and implementation of plans already completed is contingent upon fund availability. There are concerns that attention to water quality problems has peaked. Such conferences as the 1981 National Conference of Agricultural Management and Water Quality held at Iowa State University, designed to summarize achievements and chart future directions, may have been a last hurrah in a significant period of water quality research.

Solution of agricultural water quality problems has been hampered by the generally impatient nature of Americans, both individually and politically. There is a tendency to spend money to resolve problems based on the assumption that quick solutions are discovered in direct proportion to the amount of money spent. However, the technical bases must be created through multidisciplinary, comprehensive, and time-consuming research. There is the danger that significant research activities will be nearly completed, only to be scuttled because of shifts in the political winds or curtailment of funding. The need for adequate technical data and an adequate planning horizon are essential for good economic analyses. Without sufficient technical data and long enough planning horizons, adequate measurements of costs and benefits will be impossible. A positive feature of the RCWP projects is their 10-year planning period. This period is adequate for identifying problems, establishing data bases, making benchline measurements, monitoring programs, and conducting economic evaluations.

Increasing attention is being focused on whether additional soil conservation and water pollution control could be achieved through cross-compliance with USDA commodity programs (*8, 73, 75*). Concepts such as guiding Agricultural Conservation Program cost-sharing funds to soil conservation practices in problem areas and variable cost-sharing programs are being evaluated and will likely influence future activities.

Looking ahead from the perspective of the accomplishments since the early 1970s provides a basis for cautious optimism. Progress in the development of technical information and computer models has facilitated the description of physical relationships and the estimation of economic impacts of alternative policies. Future progress will depend in part upon the commitment of funds and institutional and political support to water quality goals. A continuing challenge to economists is to complete good technical analyses and to apply their results in the institutional and political arenas.

REFERENCES

1. Alt, Klaus, F., and E. O. Heady. 1977. *Economics and the environment: Impacts of erosion restraints on crop production in the Iowa River Basin.* Card Report 75. Iowa State University, Ames.
2. Alt, Klaus, F., J. A. Miranowski, and E. O. Heady. 1979. *Social costs and effectiveness of alternative nonpoint pollution control practices.* In *Best Management Practices for Agriculture and Silviculture, Proceedings of the 1978 Cornell Agricultural Waste Management Conference.* Ann Arbor Science Publishers, Inc., Ann Arbor, Michigan. pp. 321-327.
3. Bailey, G. W., and Robert R. Swank, Jr. 1983. *Modeling agricultural nonpoint source pollution: A research perspective.* In *Proceedings, National Conference on Agricultural Management and Water Quality.* Iowa State University Press, Ames.
4. Bailey, G. W., and T. E. Waddell. 1979. *Best management practices for agriculture and silviculture: An integrated overview.* In *Best Management Practices for Agriculture and Silviculture, Proceedings of the 1978 Cornell Agricultural Waste Management Conference.* Ann Arbor Science Publishers, Inc., Ann Arbor, Michigan. pp. 33-36.
5. Baker, J. L., and H. P. Johnson. 1983. *Evaluating the effectiveness of BMP's from field studies.* In *Proceedings, National Conference on Agricultural Management and Water Quality.* Iowa State University Press, Ames.
6. Baker, J. L., H. P. Johnson, M. A. Borcherding, and W. R. Payne. 1979. *Nutrient and pesticide movement from field to stream: A field study.* In R. C. Loehr, D. A. Haith, M. F. Walter, and C. S. Martin [editors] *Best Management Practices for Agriculture and Silviculture.* Ann Arbor Science Publishers, Inc., Ann Arbor, Michigan. pp. 213-244.
7. Beasley, D. B., E. J. Monke, and L. F. Huggins. 1977. *The ANSWERS model: A planning look for watershed research.* Paper No. 77-2532. American Society of Agricultural Engineers, St. Joseph, Michigan.
8. Benbrook, Charles. 1979. *Integrating soil conservation and commodity programs: A policy proposal.* Journal of Soil and Water Conservation 34(4): 160-167.
9. Boggess, W., James McGrann, Michael Boehlje, and Earl O. Heady. 1979. *Farm-level*

impacts of alternative soil loss control policies. Journal of Soil and Water Conservation 34(4): 177-183.

10. Bouwes, N. W., Sr., and Robert Schneider. 1979. *Procedures in estimating benefits of water quality change.* American Journal of Agricultural Economics 61(3): 535-539.

11. Bouwes, N. W., Sr., and R. R. Schneider. 1978. *Estimating water quality benefits.* USDA Working Paper No. 55. U.S. Department of Agriculture, Washington, D.C.

12. Brady, Nyle C., Editor. 1967. *Agriculture and the quality of our environment.* In proceedings, symposium, 133rd meeting, American Association for the Advancement of Science. AAAS Publication 85. Plimpton Press, Norwood, Massachusetts.

13. Bruce, R. R., L. A. Harper, R. A. Leonard, W. M. Snyder, and A.W.T Thomas. 1975. *A model for runoff of pesticides from small upland watersheds.* Journal of Environmental Quality 4(4): 541-548.

14. Burwell, R. E., G. E. Schuman, H. G. Heinemann, and R. G. Spomer. 1977. *Nitrogen and phosphorus movement from agricultural watersheds.* Journal of Soil and Water Conservation 32(5): 226-230.

15. Carter, Luther J. 1977. *Soil erosion: The problem persists despite the billions spent on it.* Science 196(4,288): 409-411.

16. Cassler, G. L., and J. J. Jacobs. 1975. *Economic analysis of reducing phosphorus losses from agricultural production.* In *Nitrogen and Phosphorus—Food Production, Waste and the Environment.* Ann Arbor Science Publishers, Inc., Ann Arbor, Michigan.

17. Christensen, L. A., and John A. Miranowski, editors. 1982. *Perceptions, attitudes and risk: Overlooked variables in formulating public policy on soil conservation and water quality.* ERS staff report no. AGES820129. Economic Research Service, U.S. Department of Agriculture, Washington, D.C.

18. Christensen, L. A., J. R. Trierweiler, T. J. Ulrich, and M. W. Erickson. 1981. *Managing animal wastes: Guidelines for decisionmaking.* ERS-671. Economic Research Service, U.S. Department of Agriculture, Washington, D.C.

19. Christensen, L. A., and Patricia E. Norris. 1983. *Soil conservation and water quality improvement: What farmers think.* Journal of Soil and Water Conservation 38(1): 15-20.

20. Coote, D. R., E. M. MacDonald, W. T. Dickinson, R. C. Ostry, and R. Frank. 1982. *Agriculture and water quality in the Canadian Great Lakes Basin: I. Representative agricultural watersheds.* Journal of Environmental Quality 11(3): 473-481.

21. Cosper, Harold R. 1978. *The influence of tillage systems on corn yields and soil loss in Ohio, Indiana, Illinois, and Iowa.* Economics Research Service, U.S. Department of Agriculture, Washington, D.C.

22. Cosper, Harold R. 1979. *Soil taxonomy as a guide to economic feasibility of soil tillage systems in reducing nonpoint pollution.* Economics, Statistics, and Cooperative Service, U.S. Department of Agriculture, Washington, D.C.

23. Council on Environmental Quality. 1970. *Environmental quality—1970.* Washington, D.C.

24. Council on Environmental Quality. 1976. *Environmental quality—1976.* Washington, D.C.

25. Council on Environmental Quality. 1978. *Environmental quality—1978.* Washington, D.C.

26. Council on Environmental Quality. 1979. *Environmental quality—1979.* Washington, D.C.

27. Council on Environmental Quality. 1980. *Environmental quality—1980.* Washington, D.C.

28. Crawford, N. H., and A. S. Donigian, Jr. 1973. *Pesticide transport and runoff model for agricultural lands.* EPA-660/2-74-013. U.S. Environmental Protection Agency, Athens, Georgia.

29. Crosson, Pierre. 1981. *Conservation tillage and conventional tillage: A comparative assessment.* Soil Conservation Society of America, Ankeny, Iowa.

30. Crosswhite, William M., and James Meek, editors. 1982. *Water quality monitoring*

and modeling workshop—proceedings. Staff Report No. AGES810506. Economics Research Service, U.S. Department of Agriculture, Washington, D.C.

31. Dale, J. T. 1979. *Abating agricultural pollution, part II: The Rural Clean Water Program.* Journal of Water Pollution Control Federation 51(2): 232-234.

32. Dempster, T. H., and J. H. Stierna. 1979. *Procedure for economic evaluation of best management practices.* In *Best Management Practices for Agriculture and Silviculture, Proceedings, 1978 Cornell Agricultural Waste Management Conference.* Ann Arbor Science Publishers, Inc., Ann Arbor, Michigan. pp. 383-391.

33. Donigian, A. S., D. C. Beyerlein, H. H. Davis, Jr., and N. H. Crawford. 1977. *Agricultural runoff management (ARM) model version II: Refinement and testing.* EPA 600/3-77-098. U.S. Environmental Protection Agency, Athens, Georgia.

34. Donigian, A. S., and N. H. Crawford. 1976. *Modeling nonpoint pollution from the land surface.* EPA 600/3-76-083. U.S. Environmental Protection Agency, Athens, Georgia.

35. Donigian, A. S., and N. H. Crawford. 1976. *Modeling pesticides and nutrients on agricultural lands.* EPA-600/2-76-043. U.S. Environmental Protection Agency, Athens, Georgia.

36. Donigian, A. S., and N. H. Crawford. 1977. *Simulation of nutrient loadings in surface runoff with the NPS model.* EPA 600/3-77-065. U.S Environmental Protection Agency, Athens, Georgia.

37. Dyke, Paul T., and Jimmy R. Williams. 1982. *Erosion productivity impact calculator (EPIC): A demonstration of process modeling for policy analysis.* NRED Working Paper No. 3. Economics Research Service, U.S. Department of Agriculture, Washington, D.C.

38. Ellis, B. G., A. E. Erickson, A. R. Wolcott, M. Zabik, and R. Leavitt. 1977. *Pesticide runoff losses from small watersheds in Great Lakes Basin.* EPA 600/3-77-112. U.S. Environmental Protection Agency, Athens, Georgia.

39. Ellis, B. G., A. E. Erickson, and A. R. Wolcott. 1978. *Nitrate and phosphorus runoff losses from small watersheds in Great Lakes Basin.* EPA 600/3-78-028. U.S. Environmental Protection Agency, Athens, Georgia.

40. Frere, M. H., C. A. Onstad, and H. N. Holtan. 1975. *ACTMO, an agricultural chemical transport model.* U.S. Department of Agriculture—Agricultural Research Service, Washington, D.C.

41. Gianessi, Leonard P., and Henry M. Peskin. 1981. *Analysis of national water pollution control policies: 2. Agricultural sediment control.* Water Resources Research 17(4): 803-821.

42. Gianessi, Leonard P., Henry M. Peskin, and G. K. Young. 1981. *Analysis of national water pollution control policies: 1. A national network model.* Water Resources Research 17(4): 796-801.

43. Gum, Russell B., and Erick B. Oswald. 1982. *Models for evaluation of water quality improvements.* In *Water Quality Modeling and Monitoring Workshop Proceedings.* ERS staff report no. AGES810506. Economics Research Service, U.S. Department of Agriculture, Washington, D.C. pp. 219-231.

44. Gum, Russell B., Louis B. Arthur, Erick B. Oswald, and William E. Martin. 1983. *Quantified evaluation for decisions: Measuring achievement of environmental goals and analyzing trade-offs.* Agricultural Experiment Station Technical Bulletin 145. Oregon State University, Corvallis.

45. Haith, Douglas A. 1980. *Models for the analysis of agricultural nonpoint source pollution.* Draft prepared for the International Institute for Applied Systems Analysis, Luxembourg, Austria. Department of Agricultural Engineering, Cornell University, Ithaca, New York.

46. Haith, Douglas A., and L. J. Tubbs. 1981. *Operational methods for analysis of agricultural nonpoint source pollution.* Cornell Agricultural Experiment Station, SUNY at Cornell University, Ithaca, New York.

47. Haith, Douglas A., and Raymond C. Loehr, editors. 1979. *Effectiveness of soil and water conservation practices for pollution control.* EPA-600/3-79-106. U.S. Environ-

mental Protection Agency, Athens, Georgia.
48. Heady, E. O., and G. Vocke. 1982. *Programmed impacts of environmental restraints applied to U.S. agriculture.* Journal of Environmental Economics and Management 9: 143-149.
49. Heimlich, Ralph E., and Clayton W. Ogg. 1982. *Evaluation of soil-erosion and pesticide-exposure control strategies.* Journal of Environmental Economics and Management, Volume 9. pp. 279-288.
50. Hoover, Herbert, and Marc Wiitala. 1980. *Operator and landlord participation in soil erosion control in the Maple Creek Watershed in northeast Nebraska.* ESCS Staff Report NRED 80-1. Economics, Statistics, and Cooperatives Service, U.S. Department of Agriculture, Washington, D.C.
51. Indiana Heartland Coordinating Commission. 1981. *Indiana Heartland Model Implementation Project, status report.* Indianapolis.
52. Jacobs, J. J., and J. F. Timmons. 1974. *An economic analysis of agricultural land use practices to control water quality.* Journal of Agricultural Economics 56(4): 791-798.
53. Johnson, H. P., and J. L. Baker. 1981. *Development and testing of mathematical models as management tools for agricultural pollution control, volume I (summary) and volume II (data).* U.S. Environmental Protection Agency, Athens, Georgia.
54. Kasal, James. 1976. *Trade-offs between farm income and selected environmental indicators.* Technical Bulletin No. 1,550. Economics Research Service, U.S. Department of Agriculture, Washington, D.C.
55. Knisel, Walter G., editor. 1980. *CREAMS: A field scale model for Chemicals, Runoff and Erosion from Agricultural Management Systems.* Conservation Research Report No. 26. U.S. Department of Agriculture, Washington, D.C.
56. Knisel, Walter G., R. A. Leonard, and E. B. Oswald. 1982. *Nonpoint-source pollution control: A resource conservation perspective.* Journal of Soil and Water Conservation 37(4): 196-199.
57. Korsching, Peter F., and Peter J. Nowak. 1980. *Sociological factors in the adoption of best management practices. Annual Report to the Environmental Protection Agency.* Department of Sociology and Anthropology, Iowa State University, Ames.
58. Laflen, John M. 1983. *Conference summary and research needs: An agricultural perspective.* In *Proceedings, National Conference on Agricultural Management and Water Quality.* Iowa State University Press, Ames.
59. Lake, James, and James Morrison. 1977. *Environmental impact of land use on water quality. Final report on the Black Creek Project (summary).* EPA-905/9-77-007-A. Environmental Protection Agency, Washington, D.C.
60. Massey, Dean T. 1980. *The model implementation program—A cooperative effort by USDA and EPA for water quality management: An overview.* ESCS Staff Report, NRED 80-13. U.S. Department of Agriculture, Washington, D.C.
61. Massey, Dean T. 1983. *The model implementation program—A cooperative effort by USDA and EPA for water quality management.* Report in process. Economics Research Service, U.S. Department of Agriculture, Washington, D.C.
62. McElroy, A. D., S. Y. Chiu, J. W. Nebgen, A. Aleti, and F. W. Bennett. 1976. *Loading functions for assessment of water pollution from nonpoint sources.* EPA-600/2-76-151. U.S. Environmental Protection Agency, Athens, Georgia.
63. McGrann, J. M., and J. Meyer. 1979. *Farm-level economic evaluation of erosion control and reduced chemical use in Iowa.* In *Best Management Practices for Agriculture and Silviculture, Proceedings, 1978 Cornell Agricultural Waste Management Conference.* Ann Arbor Science Publishers, Inc., Ann Arbor, Michigan. pp. 359-372.
64. McKusick, Robert B. 1982. *Water quality modeling—economics components and evaluation.* In *Water Quality Modeling and Monitoring Workshop Proceedings.* ERS Staff Report No. AGES810506. Economic Research Service, U.S. Department of Agriculture, Washington, D.C. pp. 72-76.
65. Meta Systems, Inc. 1979. *Costs and water quality impacts of reducing agricultural nonpoint source pollution, an analysis methodology.* EPA-600/5-79-009. U.S. Environmental Protection Agency. Athens, Georgia.

66. Miller, William L., and Joseph H. Gill. 1976. *Equity considerations in controlling nonpoint pollution from agricultural sources.* Water Resource Bulletin 12(2): 253-261.
67. Miranowski, John, and Klaus Alt. 1978. *Policy costs in the theory of externalities and the selection of nonpoint pollution policy.* Journal of Economics 4(4): 159-162.
68. Miranowski, John, and Klaus Alt. 1983. *Best management practice implementation economics and farmer decision making.* In *Proceedings, National Conference on Agricultural Management and Water Quality.* Iowa State University Press, Ames.
69. Miranowski, John, Michael J. Monson, James S. Shortle, and Lee D. Zinser. 1982. *Final report—effect of agricultural land use practices on stream water quality: Economic analysis—15 October 1979-15 July 1981.* Department of Economics, Iowa State University, Ames.
70. Morrison, Jim. 1977. *Managing farmland to improve water quality.* Journal of Soil and Water Conservation 32(5): 205-208.
71. Nowak, Peter J., and Peter F. Korsching. 1983. *Social and institutional factors affecting the adoption and maintenance of agricultural BMPs.* In *Proceedings, National Conference on Agricultural Management and Water Quality.* Iowa State University Press, Ames.
72. Ogg, Clayton W. 1981. *Building water quality objectives into conservation planning models.* In *Proceedings, Symposium on Nonpoint Pollution Control: Tools and Techniques for the Future.* Interstate Commission on the Potomac River Basin, Rockville, Maryland. pp. 129-134.
73. Ogg, Clayton W., Arnold B. Miller, Kenneth C. Clayton, and James D. Johnson. 1982. *Soil conservation under integrated farm programs.* American Journal of Agricultural Economics 64(5): 1,085.
74. Ogg, Clayton W., Harry B. Pionke, and Ralph E. Heimlich. 1983. *A linear programming economic analysis of lake quality improvement using phosphorus buffer curves.* Journal of Water Resources Research 19(1): 21-31.
75. Ogg, Clayton W., James D. Johnson, and Kenneth C. Clayton. 1982. *A policy option for targeting soil conservation expenditures.* Journal of Soil and Water Conservation 37(2): 68-72.
76. Ogg, Clayton W., Lee A. Christensen, and Ralph E. Heimlich. 1979. *Economics of water quality in agriculture—a literature review.* ESCS 58. Economics, Statistics, and Cooperative Service, U.S. Department of Agriculture, Washington, D.C.
77. Omernik, J. M. 1977. *Nonpoint source—stream nutrient level relationships: A nationwide study.* EPA 600/3-77-105. Corvallis Environmental Research Laboratory, Corvallis, Oregon.
78. Onishi, H., and E. R. Swanson. 1974. *Effects of nitrate and sediment constraints on economically optimal crop production.* Journal of Environmental Quality 3(3): 234-238.
79. Osteen, Craig, and Wesley D. Seitz. 1978. *Regional economic impacts of policies to control erosion and sedimentation in Illinois and other Corn Belt states.* American Journal of Agricultural Economics 60(3): 510-517.
80. Osteen, Craig, W. W. Seitz, and J. B. Stall. 1980. *Toward instream water quality management.* Contract report prepared for the Environmental Research Laboratory. U.S. Environmental Protection Agency, Athens, Georgia.
81. Oswald, Eric B. 1978. *The surface water quality impacts of resource management plans: A structure for analysis.* Economic Research Service, U.S. Department of Agriculture, Washington, D.C.
82. Reneau, D. R., and C. R. Taylor. 1979. *An economic analysis of erosion control options in Texas.* In *Best Management Practices for Agriculture and Silviculture, Proceedings, 1978 Cornell Agricultural Waste Management Conference.* Ann Arbor Science Publishers, Inc., Ann Arbor, Michigan. pp. 393-418.
83. Robillard, P. D., M. F. Walter, and R. Gilmour. 1979. *Development of BMPs for agriculture—New York State strategy.* In *Best Management Practices for Agriculture and Silviculture, Proceedings of the 1978 Cornell Agricultural Waste Management Conference.* Ann Arbor Science Publishers, Inc., Ann Arbor, Michigan. pp. 581-595.

84. Robillard, P. D., and Michael F. Walter. 1983. *A framework for selecting agricultural nonpoint source controls.* In *Proceedings, National Conference on Agricultural Management and Water Quality.* Iowa State University Press, Ames.

85. Robillard, P. D., Michael F. Walter, and Roger W. Hexem. 1980. *Evaluation of agricultural sediment control practices relative to water quality planning.* Journal of the Northeastern Agricultural Economics Council IX(1): 29-36.

86. Seitz, W. D., C. Osteen, and M. C. Nelson. 1979. *Economic impacts of policies to control erosion and sedimentation in Illinois and other Corn Belt states.* In *Best Management Practices for Agriculture and Silviculture, Proceedings, 1978 Cornell Agricultural Waste Management Conference.* Ann Arbor Science Publishers, Inc., Ann Arbor, Michigan. pp. 373-383.

87. Seitz, W. D., D. M. Gardner, and J. C. VanEs. 1980. *Recognizing farmers attitudes and implementing nonpoint source pollution control policies.* University of Illinois contract report. U.S. Environmental Protection Agency, Athens, Georgia.

88. Seitz, W. D., D. M. Gardner, S. K. Gove, K. L. Guntermann, et al. 1978. *Alternative policies for controlling nonpoint agricultural sources of water pollution.* EPA 600/5-78-005. U.S. Environmental Protection Agency, Athens, Georgia.

89. Seitz, W. D., and Earl R. Swanson. 1980. *Economics of soil conservation from the farmer's perspective.* American Journal of Agricultural Economics 62(5): 1,084-1,088.

90. Sharp, B.M.H., and S. J. Berkowitz. 1979. *Economic, institutional and water quality considerations in the analysis of sediment control alternatives: A case study.* In *Best Management Practices for Agriculture and Silviculture, Proceedings, 1978 Cornell Agricultural Waste Management Conference.* Ann Arbor Science Publishers, Inc., Ann Arbor, Michigan. pp. 429-453.

91. Smith, C. N., G. W. Bailey, R. A. Leonard, and G. W. Langdale. 1978. *Transport of agricultural chemicals from small upland piedmont watersheds.* EPA 600/3-78-056. U.S. Environmental Protection Agency, Athens, Georgia.

92. Smith, E. E., E. A. Lang, G. L. Casler, and R. W. Hexem. 1979. *Cost effectiveness of soil and water conservation practices for improvement of water quality.* In D. A. Haith and R. C. Loehr [editors] *Effectiveness of Soil and Water Conservation Practices for Pollution Control.* EPA-600/3-79-106. U.S. Environmental Protection Agency, Athens, Georgia. pp. 147-205.

93. Smathers, Webb M., Jr., Ivery D. Clifton, Lee A. Christensen, Fred C. White, and Wesley N. Musser. 1982. *Factors influencing agricultural nonpoint pollution control in the Altamaha River Basin of Georgia.* FS82-1. Department of Agricultural Economics, University of Georgia.

94. Soil Conservation Society of America. 1982. *Resource conservation glossary, third edition.* Ankeny, Iowa.

95. Taylor, C. R., D. R. Reneau, and B. L. Harris. 1978. *An economic analysis of erosion and sedimentation in Lavon Reservoir watershed.* Bulletin TR-88. Texas Water Resource Institute, College Station, Texas.

96. Taylor, C. R., and E. R. Swanson. 1975. *The economic impact of selected nitrogen restrictions on agriculture in Illinois and 20 other regions of the U.S.* AERR 133. University of Illinois, Urbana-Champaign.

97. Taylor, C. R., and K. K. Frohberg. 1977. *The welfare effects of erosion controls, banning pesticides and limiting fertilizer application in the Corn Belt.* American Journal of Agricultural Economics 59: 24-36.

98. Unger, Samuel G. 1977. *Environmental implications of trends in agriculture and silviculture, volume I: Trend identification and evaluation.* EPA-600/3-77-131. U.S. Environmental Protection Agency, Athens, Georgia.

99. Unger, Samuel G. 1979. *Environmental implications of trends in agriculture and silviculture, volume III: Regional crop production trends.* EPA-600/3-79-047. U.S. Environmental Protection Agency, Athens, Georgia.

100. U.S. Department of Agriculture. 1981. *Soil and Water Resources Conservation Act. 1980 appraisal, part I. Soil, water, and related resources in the United States, status,*

condition, and trends. Washington, D.C.
101. U.S. Department of Agriculture. 1981. *Soil and Water Resources Conservation Act. 1980 appraisal, part II. Soil, water, and related resources in the United States. Analysis of resource trends.* Washington, D.C.
102. U.S. Department of Agriculture. 1981. *Soil and Water Resources Conservation Act Program report and environmental impact statement.* Revised draft. Washington, D.C.
103. U.S. Department of Agriculture, Agricultural Research Service, and Office of Research and Development, Environmental Protection Agency. 1975. *Control of water pollution from cropland: volume I, a manual for guideline development.* Washington, D.C.
104. U.S. Department of Agriculture, Agricultural Research Service, and Office of Research and Development, Environmental Protection Agency. 1976. *Control of water pollution from cropland: volume II, an overview.* Washington, D.C.
105. U.S. Department of Agriculture, Agricultural Stabilization and Conservation Service. 1980. *The Rural Clean Water Program.* ASCS Handbook, 1-RCWP. Washington, D.C.
106. U.S. Department of Agriculture, Science and Education Administration, and Environmental Protection Agency, Office of Research and Development. 1979. *Animal waste utilization on cropland and pastureland.* USDA Utilization Research Report No. 6, EPA-600/2-79-059.
107. U.S. Department of Agriculture, Science and Education Administration, National Soil Erosion-Soil Productivity Research Planning Committee. 1981. *Soil erosion effects on soil productivity: A research perspective.* Journal of Soil and Water Conservation 36(2): 82-90.
108. U.S. Environmental Protection Agency. 1973. *Methods and practices for controlling water pollution from agricultural nonpoint sources.* EPA-430/9-73-015. Washington, D.C.
109. U.S. General Accounting Office. 1977. *National water quality goals cannot be attained without more attention to pollution from diffused or "nonpoint" sources.* Washington, D.C.
110. U.S. General Accounting Office. 1978. *Water quality management planning is not comprehensive and may not be effective for many years.* Washington, D.C.
111. U.S. Government Accounting Office. 1982. *Cleaning up the environment: Progress achieved but major unresolved issues remain.* GAO/CED-82-72. Washington, D.C.
112. Wade, J. C., and E. O. Heady. 1976. *A national model of sediment and water quality: Various impacts on American agriculture.* CARD Report 67. Center for Agricultural and Rural Development, Iowa State University, Ames.
113. Wade, J. C., and Earl O. Heady. 1977. *Controlling nonpoint sediment sources with cropland management: A national economic assessment.* American Journal of Agricultural Economics 59: 13-24.
114. Wade, J. C., and E. O. Heady. 1978. *Measurement of sediment control impacts on agriculture.* Water Resources Research 14(1): 1-8.
115. Wadleigh, Cecil H. 1968. *Wastes in relation to agriculture and forestry.* Miscellaneous Publication No. 1065. Agricultural Research Service, U.S. Department of Agriculture, Washington, D.C.
116. Wadleigh, Cecil H. 1971. *A primer on agricultural pollution (summary).* Journal of Soil and Water Conservation 26(2): 44-65.
117. White, G. B., and E. J. Partenheimer. 1979. *The economic implications of erosion and sedimentation control plans for selected Pennsylvania dairy farms.* In R. C. Loehr, D. A. Haith, M. F. Walter, and C. S. Martin [editors] *Best Management Practices for Agriculture and Silviculture.* Ann Arbor Science Publishers, Ann Arbor, Michigan. pp. 341-357.
118. Williams R. E., and J. E. Lake. 1979. *Conservation district involvement in 208 nonpoint source implementation.* In *Best Management Practices for Agriculture and Silviculture, Proceedings, 1978 Cornell Agricultural Waste Management Conference.*

Ann Arbor Science Publishers, Inc., Ann Arbor, Michigan. pp. 57-67.

119. Wineman, J. J., W. Walker, J. Kuhner, D. V. Smith, P. Ginberg, and S. J. Robinson. *Evaluation of controls for agricultural nonpoint source pollution.* In *Best Management Practices for Agriculture and Silviculture, Proceedings o the 1978 Cornell Agricultural Waste Management Conference.* Ann Arbor Science Publishers, Inc., Ann Arbor, Michigan. pp. 599-624.

120. Wischmeier, W. H., and D. D. Smith. 1978. *Predicting rainfall-erosion losses—A guide to conservation planning.* Agricultural Handbook No. 537. U.S. Department of Agriculture, Washington, D.C.

4

Social Impacts of Water Impoundment Projects

Ted L. Napier, Elizabeth G. Bryant, and Michael V. Carter

Social scientists have made significant contributions to natural resources research in the last decade in the field of social impact assessment. While this form of research is relatively new to environmental issues, the genesis of social impact research can be traced to early sociological studies in Europe that focused on the evaluation of the social consequences of institutionalization. The major impetus for contemporary social impact research in the United States, however, came from the environmental movement (*11, 18, 19*) in the late 1960s, which culminated in passage of the National Environmental Policy Act of 1969 (NEPA).

NEPA mandates that all federally funded projects be examined closely to avoid adverse environmental consequences. Federal agencies must prepare an environmental impact statement that documents the potential impacts of each project. Initially, the environmental impact statement was perceived as being confined to the assessment of project impacts on the physical environment, but later interpretations of NEPA expanded the scope of the environmental impact statement to include economic and sociological consequences. Today a social impact assessment is required to fulfill the intent of the law.

Because many development programs were in process when NEPA was enacted, federal agencies suddenly had to justify their projects using different criteria. To fulfill the mandate for environmental impact statements and social impact assessments, agencies appropriated considerable economic resources for evaluative research. Many social scientists from universities and private consulting firms throughout the United States, attracted by the social impact assessment funding, initiated research programs. Soon great numbers of such assessments were being produced, but many of the research efforts were poor. They contributed little to decision-making and even less to professional understanding of social processes involved in natural resources development. There were many reasons why the studies were not good, but most of the contributing factors can be subsumed under two broad categories: (1) ineptness of the researchers and (2) the embryonic state of knowledge in the field.

Many "social scientists" engaged in social impact assessment research during the early 1970s were not well trained in research methods that could be used for making projections about future impacts. Few knew anything about the potential adverse social consequences of planned environmental change. Quite often it appeared that the researchers were more concerned about satisfying contractual agreements than producing useful information for decision-making.

The situation is not much better today. Even a cursory examination of numerous social impact assessment reports produced under contractual arrangements in recent years shows that few of the studies have been guided by theoretical perspectives; they are descriptive in nature. Oftentimes critical social and psychosocial factors are ignored because the sponsoring agency does not wish to examine sensitive issues.

The intellectual chaos that existed in social impact research for several years was undoubtedly a partial function of the lack of research precedents and substantive knowledge bases in the field. But we are convinced that many of the relatively useless documents produced under the guise of social impact assessment research were the product of ineptness on the part of the researchers and the inability of the sponsors to provide direction as to what was needed in the planning process. Also, the sponsors often were not motivated to produce good social impact assessments because they were simply complying with mandates. The impact assessments frequently were perceived by agency personnel as being an additional component of the legislative maze through which the agency was required to pass prior to reaching a project's implementation phase. Agency personnel wished to comply as quickly and as easily as possible with the legislative mandate and proceed to implementation. Such attitudes contributed to poor research.

As a result of these and other factors, the state of the art in social impact assessment research is not as far advanced as it should be, given the extensive human and economic resources that have been allocated to such research. This does not mean that progress has not been made because some research has contributed to our knowledge base. Our purpose here is to examine some of the social science contributions to social impact assessment research in the last 15 years. The focus is exclusively on the social impacts of water impoundments. It would be impossible to examine the research literature for all natural resources development pro-grams. But the concepts and content-specific materials discussed in the context of water projects should be applicable to some degree to other natural resources development programs that generate change within local groups. Our discussion is also confined to impacts on local people[1] because regional assessments are quite commonly made prior to the ap-proval of reservoir projects. While the regional studies have value in justifying the projects on a broad geographical basis, local impacts are often obscured in regional studies due to an averaging effect. Such a situation is unfortunate because local people often must internalize a disproportionate share of the social and economic costs associated with project implementation.

Social Impact Assessment Research

Many social phenomena have been examined by social impact assess-ment researchers, but most studies can be subsumed under two broad categories, social impacts and economic impacts. While these two cate-gories of studies are not mutually exclusive, the typology facilitates dis-cussion.

Social Consequences of Reservoir Development for Local People

Social impact assessment researchers have examined many factors in the context of reservoir development, but one of the most frequent topics for investigation has been attitudes (*6, 9, 10, 15, 26, 27, 34-44, 47, 50-52, 54*). Assessments of psychosocial responses to lake construction have been defended on the grounds that attitudes reflect past or expected ex-periences with the project impacts. Attitudinal researchers argue that in-dividuals who exhibit positive perceptions do so because they anticipate positive consequences or have already received benefits from reservoir development programs. It is also expected that individuals who antici-pate internalizing costs or who have been forced to internalize costs will exhibit negative attitudes. Examination of attitudes, therefore, should provide insight into past experiences with reservoir-induced changes or with anticipated outcomes.

The social impact assessment research we reviewed revealed extensive variance and sometimes contradictory findings on attitudes exhibited by local people toward various aspects of reservoir projects and their changed communities. Thus, generalizations about attitudinal responses

---

[1]Local people refers to individuals living in close proximity to the lake project. Interaction boundaries are used to formulate the concept of "community." Thus, individuals living close to reservoir projects and participating in local interaction networks are defined as "di-rectly affected groups."

to lake projects must be formulated cautiously. It is obvious from the literature, however, that geographic factors affect attitudes. People living in arid regions perceive lake projects much more favorably than people living in regions with more abundant rainfall. This finding probably reflects a higher priority on development of additional water supplies in the more arid regions.

Another generalization that can be made about attitudes toward lake projects is that individual assessments of proposed and completed projects are strongly influenced by costs and benefits encountered by the people directly affected. If local people believe they will benefit from a proposed reservoir or previously have benefitted from a lake project, they tend to be more favorable (1, 9, 10, 15, 21, 26, 27, 30, 34-44, 47, 50, 51). These studies also show that if local people are forced to internalize costs associated with the lake project, they tend to be more negative. It must be observed, however, that positive expectations associated with proposed projects must be realized or local people will become quite negative (21). "Vested interests" thus appear quite important in determining the social impacts of reservoir construction for local people.

In Ohio, for example, vested-interest variables were the best predictors of the responses of local people to a reservoir project (43, 44). Individuals who benefitted from the project tended to be more favorable than those who did not benefit. Research in Kentucky (10, 15, 26, 27, 30, 51) demonstrated that attitudes toward various lake projects became more negative as costs increased for local people. Nearly all respondents in one study were negative toward the project studied because they perceived that few benefits would accrue to local residents (27). Little support can be expected if local people must internalize a disproportionate share of the costs, while people living outside of the group directly affected enjoy most of the benefits (9, 16, 26, 27, 34, 40, 51).

Considerable variance has also been shown to exist among local people in terms of benefits and costs expected from lake projects. Downstream farmers, businessmen and businesswomen, recreationists, and younger people tend to favor lake projects because they believe that benefits will be forthcoming to them (51). Research in Ohio confirmed the observation that some people within affected community groups will derive more benefits from lake projects than others (43, 44). Individuals who received more benefits from the lake project were more favorable toward the development activity. Other research, however, discovered relatively little variance within directly affected groups because nearly all of the respondents were expecting to receive benefits (2, 3, 6, 50). Texas data revealed that almost all respondents believed that flood control, water supply, and recreation facilities would benefit local residents (50). Similar findings were observed in Utah (2, 3) and Montana (56).

Research in Louisiana revealed that nearly all local leaders expressed the belief that the proposed lake would generate social and economic expansion in an area that was in dire need of growth stimuli (6). The leaders supported any type of development activity because the local economy was in a state of decline. The stress caused by economic hardship was perceived to be greater than any adverse social consequences that could be introduced by the lake project. It should also be noted that the largest landowner in the proposed basin area was the federal government, which suggests that local people would lose little privately owned land to the project. This study tends to support the notion that local people exhibit positive attitudes toward projects that offer potential benefits with little commitment of local resources.

The most surprising findings encountered in our literature review were from an *ex post facto* assessment of a project in Oregon (52). According to this study, the local people remained positive toward the lake project even when the short-run impacts were shown to be negative. The reason: people remained optimistic that they would benefit from the project eventually even though they had been harmed extensively in the short-run. The belief that benefits would be realized in the future provided the local group with hope they would recover their losses. How long the local people can maintain their optimism without receiving benefits is unknown, but evidence from other research suggests that people will not remain positive for an extended period unless benefits are forthcoming (21). The Oregon study was conducted shortly after the construction crews left the area, and the economic impacts of recreation had not begun (52). If the expected economic impacts from recreation do not materialize, people's attitudes toward the project will likely become much more negative.

While assessments of general attitudes toward lake projects are useful in determining how local people respond to such development efforts, a much more significant issue for researchers to explore is *why* local people react in the manner they do. This brings us back to the vested-interest perspective introduced above.

Psychosocial Costs of Reservoir Development

One of the most frequently discussed social costs associated with reservoir construction is the psychosocial trauma experienced by people directly affected when lake projects are proposed and implemented (*1, 4, 10, 15-17, 21, 26, 29, 31, 34-44, 47, 48, 51, 54*). While the sources of trauma vary from group to group, the social consequences of the anxiety and fear are often negative. This is especially true for the poor and the aged, who are often less able to protect their own interests (4). The anxiety may be exhibited in a variety of ways, such as personal estrangement, deviant behavior, and illness. Even though the worst fears associated with reservoir development are more imagined than real (*39, 44*), the expectations of "doom" operate to adversely affect individuals who believe their community will be destroyed. If people believe that many negative things will happen to them, they will exhibit behavior consistent with that expectation, even if the belief is not based in fact.

Probably the greatest fear expressed by local people during the preconstruction phases of reservoir projects is the belief that local interaction patterns will become fragmented. Reservoir construction usually results in the physical relocation of a portion of the resident population, and some of these people usually leave the affected area. While much research has documented the desire of displaced people to relocate within the impacted community (*1, 10, 13, 26, 27, 34-44, 47, 48, 54*), some individuals find it necessary to leave, which separates them from friends and family living in the affected community. These studies strongly suggest that the ability to relocate close to original homesites and near established friends is a significant factor in reducing psychosocial stress generated by displacement of the resident population.

The psychosocial trauma experienced by displaced people does not appear to be directly associated with moving from one house to another but is more closely associated with separation from family and friends. Friendships that have existed for years can be disrupted or terminated by physical displacement (*10, 13, 15-17, 26, 27, 34-36, 54*). New friendships must be established as the composition of the resident population changes via in-migration of permanent residents. Even the social relationships of local people who are not directly affected by physical relocation of homes and farms are not immune to the affects of the changes. Concerns about safety, increased traffic congestion, relocation of highways, and other factors can reduce interpersonal contacts within a community.

There is some evidence that interaction patterns change after reservoir construction. In two studies, local people not displaced by a lake project had no trouble making new friends, but a reduction in the number of contacts with family and long-established friends occurred (*13, 26*). In contrast, there was little change in interaction patterns among local people displaced by a lake project in New England (*1*). The relocation strategy used by the development agency in the latter situation, however, was much different than the techniques used in other communities. The local population was relocated *en masse* to a new site close to the reservoir, which made it possible for local people to maintain established interpersonal relationships. This study clearly supports the observation made earlier that establishment of new residences by displaced people near their original homesites can mitigate some of the adverse effects of disruption. Still other researchers have reported that social disorganization did not occur even when the affected community was disrupted extensively (34-44). The researchers suggested that social groups are much more resilient than commonly believed when subjected to significant change forces. Such research suggests that fears expressed about the fragmentation of the local social order are grossly overstated.

Closely aligned with physical displacement is the problem of locating comparable housing for people forced to move from appropriated properties (13, 29, 34). This is not surprising because rural areas often do not have surplus housing available. Also, in Kentucky, displaced people were not satisfied with the new housing when comparable dwellings were located for them (13, 29). The relocated people complained that the qualitative aspects of the surroundings were not what they desired. Displaced people believed they were entitled to the same quality housing within the same community under "just compensation" norms associated with the use of eminent domain laws. In essence, the displaced people believed they had the right to remain in the local area and maintain established social relationships even if the federal government had to pay inflated prices for surrounding properties to replace those taken for the lake project. Many people in an Ohio-West Virginia study were quite upset because the development agency did not provide them useful housing information (34). Later research revealed that attitudes toward land acquisition procedures were influenced by the provision of housing information (38, 39).

Harsh and impersonal treatment initiated by development agency personnel can produce severe psychosocial stress for local residents (34, 38, 39, 42, 47-49). Negative attitudes toward lake projects and personal stress appear to be a partial function of the intrusion of external change agents into the lives of local residents. These feelings are compounded when the change agents have the power to employ eminent domain laws to secure private properties, which often have been in the possession of the same families for generations. Not only do local people react adversely to the appropriation of private property, they also resent the use of government power to secure the lands. Local people respond negatively to agency personnel who are insensitive to human sorrow during the land acquisition phase of the project. Lack of sensitivity during the procurement of condemned properties and the relocation of the living and the dead is a major source of psychosocial pain for local people.

The social impact assessment literature is replete with documentation of negative attitudes toward implementation policies of reservoir development agencies (13, 14, 21, 29, 31, 34-44, 47, 48, 51, 54). The literature suggests that more humanistic treatment of displaced people and more care taken when cemeteries are relocated would be rewarded with more positive acceptance of the development action. For example, some research has demonstrated that attitudes toward land acquisition policies were the best predictors of attitudes toward a project (38, 39). When local people perceived the land acquisition policies and procedures to be equitable, there was a more favorable attitude toward the lake project. When perceptions of land acquisition were negative, the corresponding attitudes toward the lake project became negative. These findings suggest that procedures used in the early phases of project implementation are crucial in affecting how the lake project will be received.

Fear that "outsiders" will change the affected community can generate considerable psychosocial stress among people directly affected. Local people often express concern that the social milieu will change to the point that it will no longer be capable of satisfying the social needs of local people. Long-term residents believe that outsiders will change the social relationships of local people and modify accepted patterns of behavior via importation of different values, beliefs, attitudes, and behavioral practices. Research suggests that some of these concerns have basis in fact, while others are considerably overstated. *Ex post facto* research in Ohio, using a longitudinal research design, revealed that fragmentation of the local social order did *not* occur even when hundreds of thousands of recreationists were attracted to the community each year and the number of permanent residents doubled in 10 years (*34-44*). Other studies have also demonstrated that extensive population changes should be expected when reservoirs are constructed (*20, 52*). The population changes begin when local residents are relocated from the basin area. Subsequent changes occur when construction crews and their families arrive. The last group to have an impact on local residents is comprised of permanent in-migrants attracted by leisure-time activities. Each of these population influences has a different but significant effect on the local group.

Ohio data clearly show that most permanent in-migrants were quickly assimilated into the reformulated social networks and were rapidly accepted by the long-term residents living in the affected community (*34-44*). The recent in-migrants did *not* adversely affect the social cohesiveness of the local group.

Data from Kentucky, however, revealed that considerable psychosocial stress was introduced into displaced people's lives by "scavengers" who came from outside the affected community group. The scavengers stole many valued antiques from the unprotected homes of people who were in the process of being relocated. The scavengers, apparently believing that the homes and antiques had been abandoned, proceeded to steal and vandalize property. Lack of police protection resulted in considerable economic loss for displaced people as well as severe psychosocial trauma. The stress was greatest for the aged, who could not understand why people would take cherished family relics (*54*).

Recreationists are another source of disruption for local people (*15-17, 34-44, 48*). In fact, recreationist-induced impacts may be the most disruptive of all changes associated with reservoir development (40). Impacts by recreationists begin soon after the initiation of lake construction and often continue for decades. Physical relocation, on the other hand, is often completed within a very few months. While the trauma associated with physical relocation is not confined to the time period of actual displacement and resettlement, the probability is that affected people will never be physically displaced again and can begin the accommodation process immediately after relocation has been completed. Recreationist impacts, however, occur repeatedly and may actually increase over time.

Some of the most significant recreationist impacts are associated with interpersonal behavior and inconsiderate actions. Litter, trespass, verbal abuse, excessive alcohol consumption, speeding violations, public displays of affection, vandalism, and invasion of privacy are among the individual behaviors that local people resent and fear (17, 34-44, 48). Each of these acts serves to alienate local people from recreationists even though most local people recognize that a minority of recreationists are responsible for the acts (48).

Recreationists also disrupt the lives of local people by their very presence in the community. Traffic congestion during the recreation seasons can impede use of local highways by residents (15, 17, 21, 43, 44). Restricted access to local roads is a serious problem for farmers who must move machinery from field to field and transport their grain to market. Noise pollution (21, 43, 44) generated by recreationists using power boats and by individuals engaged in other outdoor recreation activities disrupts the tranquility of rural environments (20, 43, 44). The large number of recreationists makes it difficult to locate a place to be alone (43, 44). These factors affect the aesthetic qualities of local people's rural lifestyle and cause stress for local inhabitants.

The disruption of, or reduction in, local public services, particularly during project construction, can produce stress for people directly affected (26, 27). Access to public services, such as highways, can be disrupted for extended periods of time and create many inconveniences for local people. Access to highways by local people may also be impeded by recreationists during certain seasons. A public service that often is affected adversely as a result of lake construction is police protection, which leads to greater fear among local people. Several studies have shown that local crime rates increase when lake projects are implemented, which places an additional strain on already over-extended control agencies (17, 21, 31, 41, 44, 47, 48, 54). Considerable fear can be generated by criminal acts in the local community by outsiders who are attracted by the lake project. Fear of criminal acts can result in constrained movement within the affected community after dark. Recreationists often frighten local people unintentionally when they pass close to occupied residences while traveling from parking areas to recreation sites. Encounters with criminal acts not only affect perceptions of personal safety but also perceptions of the lake project and the type of development programs local people would like to see implemented (*43, 44*).

While negative impacts on public services have been documented repeatedly in the social impact assessment literature, some evidence suggests that certain types of public service facilities are improved by reservoir construction. In Ohio, lake construction resulted in the improvement of local highways and other service infrastructures within the impacted community (*38, 39*). The public service facilities that existed prior to lake construction were dismantled or abandoned. The newly constructed facilities are far superior. While the highways have been used extensively by recreationists and local people for several years, little deterioration has occurred, and road repair has not yet created a financial burden for local residents. The highways built during the construction phase of the project are still in excellent condition.

In contrast, researchers observed rapid deterioration of highways near a large project in Illinois, and they documented the political rebellion of local people when it became necessary to finance repairs (*21*). Local people had become so stressed by outsiders that they decided not to pay for road repairs in hopes the poor roads would discourage use by recreationists.

Research in Oregon revealed that the expansion of public services to accommodate resident construction workers caused considerable stress among local people once the workers left (*52*). The problem was not deterioration of quality but excess capacity. Local people became concerned about how they would pay for the expanded public services that suddenly were not fully used.

A factor of considerable importance in the explanation of psychosocial stress associated with lake projects is the lack of definitive time schedules for project implementation and completion. The social impact assessment literature is quite consistent on this issue (*15, 29, 31, 34, 37-39, 51*), showing that local people are frequently stressed by the uncertainties associated with the authorization and implementation of water projects.

Development agencies could significantly reduce psychosocial stress by informing local people when the project will be initiated, how much land will be acquired, and which parcels of land will be taken. Without having this information, local people are not able to make future plans. Research indicates that definitive time tables associated with forced movement are preferable and more easily accepted by local people than uncertainties about the status of the project. Water resources agencies should establish firm time frames and abide by them when introducing a lake project.

Economic Benefits of Reservoir Development

Social impact assessment research strongly suggests that the type and magnitude of economic benefits generated by reservoir development are important in determining how local people will perceive the project. Many local residents often believe that proposed reservoir projects will be a panacea for existing problems of socioeconomic growth, and they strongly support such development efforts (*6, 12, 50, 52*). Other people within the affected communities believe that lake projects will not produce economic growth (*9, 10, 21, 26, 27, 30, 34, 36, 51*) and tend to oppose the projects. *Ex post facto* assessments of reservoir impacts usually show that the economic consequences of lake projects for local people are quite varied and sometimes negative. Several studies, for example, have demonstrated that few economic benefits accrue to local people except increases in property values (*6, 14, 15, 16, 21, 24, 25, 28, 34-44*).

Inflation in local property values creates both positive and negative impacts for local people. The increases in property values undoubtedly benefit local landowners. Some landowners become wealthy as a result of windfall profits. Land previously valued for its agricultural use often is reassessed for residential or leisure-oriented uses, which are valued at much higher prices. Increases in property values can also have positive effects on local people in terms of financing existing or expanded community services. On the other hand, because publicly owned land is not subject to taxation via intergovernmental agreements, local tax revenues can be reduced substantially as a result of taxable properties being removed from tax rolls. The adverse effects of tax revenue losses due to appropriation of private properties for lake development can be negated, however, by increases in local property values. In a Kentucky study, for example, adverse impacts of reservoir development on funding for local school systems and other public services, though expected, did not materialize because local land values increased and subsequently generated larger tax revenues than projected (*5*). We should note too that federal agencies engaged in reservoir development must subsidize local school systems for a period of time to compensate for loss of local tax revenues, and it is likely that the Kentucky findings in part reflect this economic support, at least during the initial period of adjustment to the tax losses.

The negative effects of inflated land values are primarily confined to the physically displaced members of the community. Because most displaced people prefer to relocate within the affected community's boundaries, they must pay the inflated prices for available homesites and existing housing. Such payments frequently place displaced people in an economically disadvantaged position. While displaced people may receive "fair market value" for appropriated properties, the increases in property values within the affected community serve to force many displaced people to leave the area. The problem of securing comparable properties to those taken by the state is compounded for displaced farmers who require large tracts of land to remain in production agriculture.

While the impacts of inflated property values are easily identified, resolution of the problem is more complex. Local people interpret "just compensation" to mean they will receive a fair market price for their appropriated lands that is adequate to permit them to relocate within the affected community. Unfortunately, fair market values of appropriated lands are assessed prior to the inflation of surrounding properties and in terms of use at the time of the assessment, which means that compensation for properties taken by the project may not be adequate to purchase comparable properties within the affected community. Because agricultural lands are not assessed as high as residential lands or lands in other uses, farmers particularly will be harmed by reservoir development. The adverse effects of inflation on surrounding properties is further complicated when the development agency attempts to secure condemned properties below market prices (14, 26). Such a situation compounds the economic suffering of the people directly affected. Given these circumstances, social impact assessment research shows that many people feel lake projects create considerable economic problems for them (1, 14, 26).

Reservoir construction can be an important stimulus for expansion of local businesses (20, 52). Imported construction workers and their families require a wide range of goods and services. Housing needs of construction workers must also be met, which often creates an expansion in the local building industry because available housing is often lacking in rural communities (20).

The economic benefits of reservoir development for local people may be short-lived due to outmigration of construction workers when the project is completed (52). The economic "boom" and "bust" experienced by many reservoir-impacted groups is not inevitable, however. Other than minor traffic congestion because of commuting construction workers, there usually are few adverse economic impacts on directly affected communities when construction workers live outside the communities being affected by lake projects (5, 39). Among the reasons: Local people are not required to provide expanded public services or to incorporate the workers and their families into reformulated social groups.

An analysis of national data on in-migration of construction workers found that about 30 percent of all lake construction workers come from communities outside the affected area (20). Most of the imported people are highly skilled personnel who locate relatively close to the construction site so they have ready access to it. The management-type personnel imported most frequently are employed to implement a project, and they do so as quickly as possible. They are more concerned about meeting contractual agreements than about humanistic considerations that bear on a project's acceptance locally. This impersonal orientation of management personnel may explain the reaction of local people to the harsh, impersonal treatment by construction crews and land acquisition agents noted earlier.

Ex post facto assessments of economic impacts often demonstrate that expected benefits associated with lake projects are not achieved. Two important explanatory factors are overestimation of recreation user days once the lake has been created and an inability to predict the magnitude of expenditures made in the local community by the recreationists who visit the facility. Outdoor recreation development generates few economic benefits for local people (*7, 8, 53, 57*). When lake projects are located near urban communities and visited primarily by day users, the economic impacts on local community groups are almost certain to be small. Recreationists usually bring their consumable goods with them and make few purchases in the local community. If recreationists do not make purchases in the local community, then local economies cannot benefit from increased recreation activities.

A major exception to the research noted above was an investigation of a lake project in the Ozarks (*22*). One possible explanation for the economic benefits received by local people in this particular study is the lake's location. The structure is many miles from an urban area, which means that recreationists are forced by circumstances to make purchases in local communities.

Another factor that contributes to unrealistic expectations of local economic growth is the overestimation of recreation user days made during a project's planning phases. Many times the estimated user days are never achieved. For example, local economic benefits were not achieved in an Illinois project because the number of recreationists never reached the projected level (*23, 24*). Evaluation of sales within the project's impacted area revealed some increases in retail sales in two of the larger towns located near the reservoir, but actual sales were not of the magnitude expected. Practically no changes in retail sales were observed in the other towns located in the impacted area (*33*). What small increases in economic activity occurred no doubt were negated by investments in public services (*21, 33*).

The regional variations associated with social impact research are easily observed in *ex post facto* assessments of economic impacts. Expanded supplies of water for irrigation were perceived by Utah residents to benefit nearly everyone in the area (*2, 3*). Agricultural production increased about 26 percent as a result of the irrigation water becoming available. Similar results were found in Texas, where expanded water supplies resulted in an expansion of agricultural activity, primarily cattle production (*55*). This was contrary to predictions that row-crop production would expand because of flood control and water supplies being made available by the lake project.

Supporting the Texas study results was an examination of 60 small watersheds in the Southeast, the Missouri Valley, and the Mississippi Valley (32). This examination demonstrated that flood control did not result in major expansion of row-crop production in the floodplain. The lack of row crops in the floodplain was attributed to the prohibitive costs of putting marginal land into agricultural production. More of the floodplain was used for unimproved pasture and forests.

All of these findings (2, 3, 32, 55) question the belief that reservoir development will result in land use shifts to agricultural production. Factors other than safety from flooding are involved in agricultural production decisions (32). Construction of a reservoir simply offers the farmer who owns floodplain land an option he or she did not have previously.

While reservoir development does not enhance row-crop agriculture in the protected floodplain, studies have documented other land use changes with significant economic implications. In one project area, land use has been and is changing in the impacted community from production agriculture to residential uses (43, 44). The result is higher land values and out-migration of farmers. Farming could soon cease to be a viable part of in the economic infrastructure of the affected area (42, 43).

Significant land use changes were observed in a Kentucky project area also, where development of recreation cottages immediately began to transform local land use patterns near a major reservoir (46). The land use shift in this case was from agricultural and extractive industries to leisure-oriented economic activities. Also observed was the fact that the settlement patterns around the lake will create severe economic problems for the local group when public services must be provided in the future. The same problem was recognized in Ohio, where researchers posited that provision of central water and sewage treatment facilities will be extremely costly in the future as a result of strip-development (44). An Illinois study observed a shift in land use from agricultural to recreational, but noted little residential development. Given the fact that recreational activities usually do not produce extensive economic growth, the loss of agricultural production probably will not be compensated for by recreation spending. Thus, the economic studies we reviewed strongly suggest that reservoir development does not necessarily result in economic expansion. In fact, the research suggests that local groups probably are harmed to some extent economically when all costs are considered.

Accomplishments of Social Impact Assessment Researchers

The social impact assessment research we reviewed demonstrates that reservoir projects generate extensive changes within the community groups directly affected. The studies also show that social scientists have made significant contributions to existing knowledge bases in the field of social impact assessment. Social scientists have examined many important issues while conducting social impact assessment research and have contributed substantially to people's understanding of the social processes that operate within groups directly affected by reservoir projects. Among the most important sociological issues investigated by contemporary researchers interested in reservoir development are the psychosocial stress created when physical displacement separates friends and family, the anxiety generated by the intrusion of "outsiders" into the lives of local people, the consequences of population changes produced by construction workers and by permanent in-migrants, the psychosocial stress produced by visitors to lake projects, the anxiety generated among local residents when land acquisition agents employ harsh treatment, the disruption of local public services by project construction, the expanded demands made on public services by construction workers and recreationists, the increased crime rates within impacted communities and the corresponding decline in the feelings of personal safety of local residents, the disruption of the tranquility in the rural setting by construction activity and recreation, the disappointments associated with expectations of socioeconomic growth that are never realized, and the anxiety generated among local people by the lack of definitive timetables for project implementation. Efforts that contribute to resolving these problems are important research endeavors.

Some of the most important economic issues examined by researchers interested in reservoir impact assessment include the consequences of increased local property values, the impacts of inadequate compensation for properties appropriated by the development agency, the problems associated with financing expanded public services, the economic problems created by "boom" and "bust" situations, the distributional aspects of expenditures made by recreationists in local communities, the changes in economic activity in the restructured community, the economic consequences of land use changes in the area surrounding the lake project and in the protected floodplain, and the problems associated with financing future services within communities that do not control the patterns of residential development. These topics are relevant to problem resolution and are also worthy research issues.

The social impact assessment research reviewed here indicates that local groups are affected significantly by reservoir development and that considerable variance exists in terms of the local group's reactions to the changes. An important finding produced by researchers is that anticipated consequences of reservoir development are often never realized. Local people who are convinced that significant expansion of the socioeconomic infrastructure will occur are probably destined to be disappointed.

Individuals who fear the local social milieu will be destroyed are probably creating their own purgatory because the evidence suggests that severe adverse consequences will not be experienced. The evidence strongly suggests that social groups can adapt to extensive changes and are able to accommodate many types of changes introduced by reservoir development. But such an interpretation should not be construed to suggest that adverse social impacts should be ignored. Each negative impact should be carefully studied, plans to solve the problem should be conceived, and corrective action should be taken. It is unfair for an institutional representative of society to expect local people to sacrifice for the common good when it is not necssary for them to do so. Equity issues demand that all adverse impacts should be mitigated within reason.

All of the issues we have noted can be solved by carefully conceived action. Most of the problems identified in the literature are associated with the implementation procedures employed by the development agency, the lack of future planning by the local community group, or ignorance of actual reservoir impacts by local people and agency personnel. To successfully solve the problems identified by social impact assessment researchers will necessitate that local people and agency personnel work together. There is no longer any excuse for either group to operate independently. Research data strongly suggest that knowledge possessed by one of the groups is essential for the other to fulfill its role in the planning process.

Social scientists who are competent in social impact assessment could effectively serve as consultants in the decision-making process by providing information to both parties concerning important social issues. Even with the limited knowledge base that presently exists, much can be said about the consequences of reservoir development.

The major obstacle to cooperative effort is trust. Local people usually do not put much confidence in agency personnel, and agency people often view the involvement of local people in the planning process with disdain. These attitudes must be changed if cooperation is to be achieved. Ironically, the major impediment to good planning is the very issue that must be resolved before cooperation can be achieved. Until the parties in the planning process become committed to cooperative efforts, an adversary role will always be assumed when reservoir projects are proposed.

When the two parties are engaged in conflict, the role of social scientists changes drastically. Scientists may align themselves with one of the combatants and become advocates or conduct research and publish results in the standard professional outlets. In either event the utility of the scientist in the planning process is seriously curtailed.

REFERENCES

1. Adler, Steven P., and Edmund F. Jansen. 1978. *Hill reestablishment: Retrospective community study of a relocated New England town.* Institute for Water Resources, Fort Belvoir, Virginia.
2. Andrews, Wade H., Alten B. Davis, Kenneth S. Lyon, Gary E. Madsen, R. Welling Roskelley, and Bruce L. Bower. 1972. *Identification and measurement of quality of life elements in planning for water resources development: An exploratory study.* Research Report 2. Institute for Social Science Research on Natural Resources, Utah State University, Logan.
3. Andrews, Wade H., G. E. Madsen, and G. J. Legaz. 1974. *Social impacts of water resources developments and their implications for urban and rural development: A post-audit analysis of the Weber Basin Project in Utah.* Research Report 4. Institute for Social Science Research on Natural Resources, Utah State University, Logan.
4. Baskin, John. 1976. *New Burlington: The life and death of an American village.* W. W. Norton and Company, Inc., New York, N.Y.
5. Bates, Clyde T. 1969. *The effect of a large reservoir on local government revenue and expenditure.* Research Report 23. Water Resources Institute, University of Kentucky, Lexington.
6. Bertrand, Alvin L. 1975. *Middle Fork Bayou D'Arbonne Reservoir Project: A feasibility and social impact assessment study.* Louisiana State University, Baton Rouge.
7. Bertrand, Alvin L., and James G. Hoover. 1973. *Toledo Bend Reservoir: A study of uses, characteristics, patterns and preferences.* Bulletin 675. Louisiana State University and Agricultural and Mechanical College, Baton Rouge.
8. Bryant, Elizabeth G., and Ted L. Napier. 1981. *The socio-economic impact of outdoor recreation development: A literature review.* In Ted L. Napier [editor] *Outdoor Recreation Planning, Perspectives and Research.* Kendall/Hunt Publishing Company, Dubuque, Iowa. pp. 103-112.
9. Bultena, Gordon. 1975. *Community values and collective action in reservoir development.* Water Resources Research Institute, Iowa State University, Ames.
10. Burdge, Rabel J., and Richard L. Ludtke. 1970. *Factors affecting relocation in response to reservoir development.* Research Report 29. Water Resources Institute, University of Kentucky, Lexington.
11. Catton, William R., Jr., and Riley E. Dunlap. 1980. *A new ecological paradigm for post-exuberant sociology.* American Behavioral Scientist 24(1): 15-47.
12. Cook, Earl, Ruth Schaeffer, James Stribling, Dwayne Baumann, and W. Simkowski. 1974. *Reservoir impact study.* Water Resources Institute, Texas A&M University, College Station.
13. Donnermeyer, Joseph F., and Peter F. Korsching. 1976. *Long-term and age-related adjustment problems associated with forced relocation in eminent domain cases.* In *Sociology in the South.* Southern Association of Agricultural Scientists.
14. Donnermeyer, Joseph F., Peter F. Korsching, and Rabel J. Burdge. 1974. *An interpretative analysis of family and individual economic costs due to water resource development.* Water Resources Bulletin 10(1): 91-100.
15. Drucker, Phillip, Charles R. Smith, and Allen C. Turner. 1972. *Impact of a proposed reservoir on local land values.* Research Report 51. Water Resources Institute, University of Kentucky, Lexington.
16. Drucker, Phillip, Charles Smith, and Edward B. Reeves. 1974. *Displacement of persons by major public works.* Research Report 80. Water Resources Institute, University of Kentucky, Lexington.
17. Drucker, Phillip, Jerry E. Clark, and Lesker D. Smith. 1973. *Sociocultural impact of reservoirs on local government institutions: Anthropological analysis of social and cultural benefits and costs from stream control measures—phase 4.* Research Report 65. Water Resources Institute, University of Kentucky, Lexington.
18. Dunlap, Riley E., and William R. Catton, Jr. 1979. *Environmental sociology.* Annual Review of Sociology 5: 243-273.

19. Dunlap, Riley E., and William R. Catton, Jr. 1979. *Environmental sociology: A framework for analysis.* In T. O'Riordon and R. C. d'Arge [editors] *Progress in Resource Management and Environmental Planning.* John Wiley and Sons, New York, N.Y. pp. 57-85.
20. Dunning, C. Mark. 1982. *Construction work force characteristics and community service impact assessment.* Water Resources Bulletin 18(2): 239-244.
21. Dwyer, John F., Robert D. Espeseth, and David L. McLaughlin. 1981. *Expected and actual local impacts of reservoir recreation.* In Ted L. Napier [editor] *Outdoor Recreation Planning, Perspectives and Research.* Kendall/Hunt Publishing Company, Dubuque, Iowa. pp. 113-120.
22. Garbacz, Charles. 1971. *The Ozarks: Recreation and economic development.* Land Economics 47: 418-421.
23. Gramann, James H. 1981. *An examination of visitation patterns to Lake Shelbyville.* In Rabel J. Burdge and Paul Opryszek [editors] *Coping with Change: An Interdisciplinary Assessment of Lake Shelbyville Reservoir.* Institute for Environmental Studies, University of Illinois, Urbana. pp. 269-281.
24. Gramann, James H. 1981. *Local economic impacts of visitor expenditures for recreation.* In Rabel J. Burdge and Paul Opryszek [editors] *Coping with Change: An Interdisciplinary Assessment of the Lake Shelbyville Reservoir.* Institute for Environmental Studies, University of Illinois, Urbana. pp. 282-298.
25. Hargrove, M. B. 1971. *Economic development of areas contiguous to multi-purpose reservoirs: The Kentucky-Tennessee experience.* Research Report 21. Water Resources Institute, University of Kentucky, Lexington.
26. Johnson, Sue, and Rable J. Burdge. 1974. *An analysis of community and individual reactions to forced migration due to reservoir construction.* In D. R. Field, J. C. Barron, and B. F. Long [editors] *Water and Community Development: Social and Economic Perspectives.* Ann Arbor Science Publishers, Inc., Ann Arbor, Michigan. pp. 169-188.
27. Johnson, Sue, Rable J. Burdge, and William F. Schweri. 1976. *Report of household survey—Red River residents due for relocation.* Center for Development Change, University of Kentucky, Lexington.
28. Kasal, James. 1978. *Effects of small watershed development on land values.* Agricultural Economic Report 404. U.S. Department of Agriculture, Washington, D.C.
29. Korsching, Peter F., Joseph F. Donnermeyer, and Rabel J. Burdge. 1980. *Perception of property settlement payments and replacement housing among displaced persons.* Human Organization 39(4): 332-338.
30. Ludtke, Richard L., and Rabel J. Burdge. 1970. *Evaluation of the social impact of reservoir construction on the residential plans of displaced persons in Kentucky and Ohio.* Research Report 26. Water Resources Institute, University of Kentucky, Lexington.
31. Mack, Ruth. 1974. *Criteria for evaluation of social impact of flood management alternatives.* Institute of Public Administration, New York, N.Y.
32. Mattson, C. Dudley. 1975. *Effect of the small watershed program on major uses of land: Examination of 60 projects in the Southeast, Mississippi Delta, and Missouri River tributaries regions.* Agricultural Economic Report 279. Economic Research Service, U.S. Department of Agriculture, Washington, D.C.
33. McLaughlin, David, and Nuntana Suwanamalik. 1981. *Changes in business and government expenditures.* In Rabel J. Burdge and Paul Opryszek [editors] *Coping with Change: An Interdisciplinary Assessment of the Lake Shelbyville Reservoir.* Institute for Environmental Studies, University of Illinois, Urbana. pp. 159-181.
34. Napier, Ted L. 1971. *The impact of water resources development on local rural communities: Adjustment factors to rapid change.* Ph.D. dissertation. Department of Sociology, Ohio State University, Columbus.
35. Napier, Ted L. 1972. *Social-psychological response to forced relocation due to watershed development.* Water Resources Bulletin 8(4): 784-794.
36. Napier, Ted L. 1974. *An analysis of the social impact of water resource development and subsequent forced relocation of population upon rural community groups: An at-*

titudinal study. Bulletin 1080. Ohio Agricultural Research and Development Center, Wooster.

37. Napier, Ted L., and Cathy J. Wright. 1976. *A longitudinal analysis of the attitudinal response of rural people to natural resources development: A case study of the impact of water resource development.* Research Bulletin 1083. Ohio Agricultural Research and Development Center, Wooster.

38. Napier, Ted L., and Cathy W. Moody. 1977. *The social impact of forced relocation of rural populations due to planned environmental modification.* Western Sociological Review 8(1): 91-104.

39. Napier, Ted L., and Cathy W. Moody. 1979. *The social impact of watershed development: A longitudinal study.* Water Resources Bulletin 15(3): 692-705.

40. Napier, Ted L., and Elizabeth G. Bryant. 1981. *Attitudes of local people toward the uses made of a multipurpose reservoir.* In Ted L. Napier [editor] *Outdoor Recreation Planning, Perspectives and Research.* Kendall/Hunt Publishing Company, Dubuque, Iowa. pp. 121-133.

41. Napier, Ted L., Elizabeth G. Bryant, and Michael Carter. 1981. *Impact of lake construction: The local perspective.* Socio-Economic Information 631. Ohio State University, Columbus.

42. Napier, Ted L., Elizabeth G. Bryant, and Steve McClaskie. 1981. *The social impact of reservoir construction in the urban fringe.* Paper presented at the 1981 Rural Sociological Society Meeting, Guelph, Canada. Economics and Sociology Occasional Paper No. 817. Department of Agricultural Economics and Rural Sociology, Ohio State University, Columbus.

43. Napier, Ted L., Michael V. Carter, and Elizabeth G. Bryant. 1982. *A test of a vested interests perspective in a reservoir impacted community.* Paper presented at the 1982 Rural Sociological Society Meeting, San Francisco, California. Economics and Sociology Occasional Paper No. 939. Department of Agricultural Economics and Rural Sociology, Ohio State University, Columbus.

44. Napier, Ted L., Michael V. Carter, and Elizabeth G. Bryant. 1982. *Local attitudes toward alternative uses of a reservoir project.* Water Resources Bulletin 18(2): 211-219.

45. Opryszek, Paul. 1981. *Changes in land use and zoning.* In Rabel J. Burdge and Paul Opryszek [editors] *Coping with Change: An Interdisciplinary Assessment of the Lake Shelbyville Reservoir.* Institute for Environmental Studies, University of Illinois, Urbana. pp. 203-213.

46. Prebble, Billy R. 1969. *Patterns of land use change around a large reservoir.* Water Resources Institute, University of Kentucky, Lexington.

47. Roper, Roy E. 1981. *Social effects of land acquisition.* In Rabel J. Burdge and Paul Opryszek [editors] *Coping with Change: An Interdisciplinary Assessment of the Lake Shelbyville Reservoir.* Institute for Environmental Studies, University of Illinois, Urbana. pp. 249-268.

48. Roper, Roy E. 1981. *Weekend residents: Taking without caring.* In Rabel J. Burdge and Paul Opryszek [editors] *Coping with Change: An Interdisciplinary Assessment of Lake Shelbyville Reservoir.* Institute for Environmental Studies, University of Illinois, Urbana. pp. 299-312.

49. Roseman, C. C., and S. M. Ives. 1981. *A survey of households relocated due to reservoir construction.* In Rabel J. Burdge and Paul Opryszek [editors] *Coping with Change: An Interdisciplinary Assessment of the Lake Shelbyville Reservoir.* Institute for Environmental Studies, University of Illinois, Urbana. pp. 238-248.

50. Singh, Raghu N., and Kenneth P. Wilkinson. 1974. *On the measurement of environmental impacts of public projects from a sociological perspective.* Water Resources Bulletin 10(3): 415-425.

51. Smith, C. R. 1970. *Anticipation of change, a socioeconomic disruption of a Kentucky county before reservoir construction.* Research Report 28. Water Resources Institute, University of Kentucky, Lexington.

52. Smith, Courtland L., Thomas C. Hogg, and Michael J. Reagan. 1971. *Economic development: Panacea and perplexity for rural areas.* Rural Sociology 36(2): 173-186.

53. Stoevener, H. H., R. B. Rettig, and S. D. Reiling. 1970. *Economic impacts of outdoor recreation: What have we learned?* In Donald R. Field, James C. Barron, and Burl F. Long [editors] *Water and Community Development: Social and Economic Perspectives.* Ann Arbor Science Publishers, Ann Arbor, Michigan. pp. 235-255.

54. Stoffle, Richard W., Charles R. Smith, Danny L. Rasch, and Anita M. Duschak. 1981. *The scavengers: An unanticipated human impact of a Kentucky dam.* Social Impact Assessment 65/66: 11-16.

55. Trock, Warren L. 1972. *An assessment of effects of small watershed development.* In *Watersheds in Transition.* Department of Agricultural Economics and Rural Sociology, Texas A&M University, College Station.

56. Tureck, Hugo. 1972. *The social impact of the Libby Dam, Lincoln County: The case of absentee or exra-local influence.* Joint Water Resources Research Center, University of Montana, Missoula.

57. Van Ness, Albert W. 1968. *Social and economic factors affecting household expenditures for outdoor recreation in northeastern South Dakota.* Masters thesis. South Dakota State University, Brookings.

II
Current and Prospective
Water/Agricultural Issues

5

The Future of Agriculture in the North Central Region

Earl R. Swanson and Earl O. Heady

Agriculture's need for and use of water in the North Central Region will depend upon food demand, agricultural productivity, and water supply variables. The food demand and agricultural productivity variables are mainly domestic but include an international component. Of course, the potential uses of water in the region will depend upon its supply in the future, both in absolute quantity and the residual remaining after higher valued, competing uses have been met, together with the costs of pumping and distributing it.

The Role of Export Demand

If U.S. agriculture operated only in a domestic market, there would probably be little economic incentive for greater water use in the North Central Region. The nation could produce a large amount of food relative to domestic demand; agricultural commodity prices would be lower than in recent years; and the demand for water in agriculture would be substantially less. Use of irrigation in the region would be minor except in the western tier of states where sustainable or rechargeable water levels would allow irrigation to be economic in terms of depth of lift and energy prices. However, U.S. agriculture is unlikely to encounter this domestically centered demand scenario because of growing export demand. The nation currently exports the product from nearly a third of its grain acreage. The export share of production in 1981 was 68 percent for wheat, 35 percent for corn, and 56 percent for soybeans. Given the rapid growth in grain exports over the past two decades and world projections of population and per capita income growth, some agricultural economists predict a further, mammoth spurt in exports by 2000; they suggest that problems of the future are more nearly those of capacity to produce than of surpluses (17, 26).

The magnitude of this export demand for U.S. agricultural commodities will also depend upon the growth of agricultural productivity in the rest of the world, especially developing countries. Our own estimates suggest that world food production could grow at an annual rate of 3.5 percent to 2000, while population growth is at the rate of 2.5 percent a year (13). Of course, growth in per capita income and greater consumption of grain and meat per capita could still cause the real price of agricultural commodities to increase. In fact, considerable increase in real prices of agricultural commodities is probably necessary to maintain irrigation at its current level in the year 2000.

The extent to which real prices will increase depends upon agricultural productivity growth in both developing and developed countries over the next two decades. Productivity growth obviously has considerable potential in developing countries if policies are used that provide adequate incentives to farmers. Currently, developing countries have about 64 percent of the world's cereal area but produce only 40 percent of the total supply. This compares with 36 percent of the area and 60 percent of the supply in developed countries. Developing countries increased per hectare yields from 1.14 tons in 1934-1938 to 1.40 tons in 1973-1975; developed countries increased yields from 1.15 tons to 3.00 tons over the same time span.

Because the location of developing countries corresponds roughly with the more tropical area of the world, these countries actually should be able to increase yields more rapidly than developed countries because of the opportunities for multiple cropping and the potential for use of water and solar energy resources throughout the year. The Food and Agriculture Organization (FAO) estimates that developing countries can increase food grain production by 3.7 percent per annum to 2000, with 26 percent of this coming from increases in land area, 14 percent from greater crop intensity, and 60 percent from higher yields on currently cropped land (10). This improved technology would come through a considerable capital investment in cultivated land and newly developed land. It is estimated that rice yields in developing countries could increase from 2.1 tons per hectare in 1979 to 3.2 tons per hectare in 2000, corn yields from 1.6 to 2.4 tons per hectare, and wheat yields from 1.4 to 2.4 tons per hectare.

According to FAO estimates, about a quarter of the world's ice-free land is potential farmland, and only about half of that in developing countries is currently being farmed (10). Some of this land, however, may represent fragile soils, while reclamation and irrigation costs may be substantial on other land. Other fractions of the land involve infrastructure investment in roads and marketing facilities. Hence, for the next 20 to 30 years at least, there is basis for optimism in developing countries if capital is available and policies are appropriate to provide the means and incentives. The more successful these countries are in increasing food supplies, the smaller will be the increase in demand for grain exports from the United States.

Of course, U.S. food exports will continue to be an important component of world food supplies. If progress in U.S. agricultural productivity over the next 20 years parallels that of the past 30 years, the United States will play an important role in augmenting world food supplies, the tendency of real prices for food to rise will be dampened, and irrigated agriculture will decline in the North Central Region. In the less likely event that U.S. agriculture should approach or reach its capacity to increase output, real prices for agricultural commodities could rise sharply and greater economic encouragement of irrigation in the North Central Region would prevail. Due to the water mining currently in the Ogallala Aquifer and the eventual need to drop water withdrawals nearer recharge rates, high real prices for food would simply cause irrigation in the western portion of the region to be maintained at a higher level than otherwise over the next two or three decades and then decline to approximate recharge rates.

Production Technology and Irrigation

If improved technology and productivity growth in all of U.S. agriculture can be maintained over the next three decades, the need to develop supplemental irrigation over the eastern Corn Belt and to maintain water irrigation at higher levels in the western Corn Belt would be considerably reduced. Other technology could thus serve as a substitute for irrigation. On the other hand, if technological improvement in U.S. agriculture rapidly approaches a limit and yields and productivity growth attain a plateau, further development and maintenance of irrigation in the region would be encouraged.

Several agricultural specialists have suggested that U.S. agricultural productivity is approaching or has reached a yield plateau. Even though the U.S. corn and soybean yields for 1950-1972 published in a report by the National Academy of Sciences indicates no levelling of yields of these crops, the report points to a number of biological limitations on crop production imposed by such factors as energy availability and the leaf area of plants that will influence future yields (21). Other researchers indicate that yield curves are beginning to flatten. One scientist even suggests that a slowdown in agricultural productivity has already occurred and that what growth is still forthcoming can be imputed to past research. Crosson (7) suggested that while yield increases had started to slacken in the mid-1970s weather also had not been as favorable as in the 1960s. Another analysis of data through the mid-1970s produced evidence of declines in the rates of yield increase (18). Still another researcher, contending that increased use of fertilizer accentuated the rate of yield increase in the period 1955-1970, believes that the yield from now to the end of the century might parallel trends over the period 1935-1955.[1] An initial assessment by Wittwer (37) concluded that crop yields have plateaued. Ruttan (24) pointed out that the rate of growth in U.S. agricultural output since 1950 has averaged a little more than 1.7 percent a year, well below the 2 to 4 percent growth rates achieved by many developed and developing countries.

Questions of Yield Plateau

If U.S. yield has in fact approached a limit due to exhausted opportunities in recent forms of technological improvements, as suggested by this group of analysts, a partial offset might be (a) a greater use of water as a substitute or (b) a smaller decline in irrigation than expected as pumping costs increase with energy prices and greater lifts.

But it is not quantitatively apparent at this time that yields have attained a plateau. Most of the above propositions were made following somewhat less favorable weather and the movement of some idle land out of supply control into production during the mid-1970s. It is likely that this land was less productive than that already in crops. Studies using more recent observations and less aggregated data found that yields have not reached a plateau (31, 32). Another study indicated that statistical proof of a developing yield plateau is not yet evident if recent observations are included (8). Soybean yields continue to increase (36). At the national level, U.S. agricultural productivity in the last 20 years shows no evidence of decline (33).

Regression studies of yield data at national, state, or county levels, or even with data from individual farms, are not apt to provide much additional insight. Improved understanding of the relative importance of the various sources of crop yield increase is more likely to come through studies of the combined effect of specific technologies, such as the recent study of corn technology (29).

While statistical verification of yield limits is not yet possible, there are some logical reasons why they could appear in the future. A major one is that fertilizer and pesticides were applied to most cropland during the

[1]Thompson, L. M. 1979. "Climate Change and World Grain Production." Unpublished manuscript prepared for the Council on Foreign Relations.

period 1950-1975, and much of this source of production increment has already been realized. Similarly, the opportunity to exploit groundwater for irrigation during that period cannot be repeated in the future. Conversely, some shift from irrigated to dryland production no doubt will occur over the next 30 years as water use must drop to recharge rates. Some agricultural specialists point to limits in use of solar energy and carbon dioxide as restraints on further, large increases in crop yields. Also, some point out that improved efficiency in converting grain into beef has been very modest, and no great breakthroughs appear imminent.

Production Potentials

There also are optimists with respect to the nation's ability to continue technological change and productivity improvement. The research institutes and colleges of the country have a much larger and more highly trained cadre of agricultural scientists than four decades ago when the tremendous surge in productivity began. While future innovations and inventions to be developed are more complex than those of the past, agricultural scientists are more highly trained—theoretically, quantitatively, and in understanding the broad interactions within the systems that represent plant and animal reproduction and growth.

The future outcome with respect to productivity growth will certainly be a function of the investment society makes in maintaining its numbers and equipment. One prediction is that if expenditures on agricultural research only offset inflation over the next two decades productivity growth will slow down to only 1 percent by 2000 (*19*). With a real growth rate of 3 percent in agricultural research, the rate averaged since World War II, the growth rate could be 1.1 percent; with a real growth rate in research and development of 7 percent, the growth rate is predicted at 1.3 percent. Fuller (*12*) expects diminishing returns from agricultural research in the future; "more and more effort will yield less and less."

It is possible to observe simultaneously that annual yield increases are, on the average, declining and to judge that substantial potentials exist for increasing yields. Using other estimating methods, some biological scientists are optimistic about the future and project huge potentials for increasing future yields. Pfeiffer (*23*) suggests that possibilities are considerable because the ratio of highest yield ever recorded to average U.S. yield in 1975 was 3.9 for corn, 7.0 for wheat, 4.0 for soybeans, 6.5 for sorghum, 5.8 for rice, 6.2 for oats, 4.8 for barley, and 3.5 for potatoes. The highest commercial 1975-1976 yields attained in the U.S. (with the highest yield ever recorded in parentheses) were 225 (338) bushels per acre for corn, 100 (216) for wheat, 65 (110) for soybeans, 250 (320) for sorghum, 150 (296) for oats, and 150 (212) for barley (*37*). Worldwide, crop yields are only a third to a half of the levels that seem reasonable and possible, while U.S. corn yields of 900 bushels per acre appear feasible (37).

There is a large genetic potential for crop yields that is unrealized because of the need for better adaptation of plants to the environments in which they are grown; recent scientific advances could result in large gains in crop productivity (4). Other evidence of potential biological capacity lies in the fact that, while typical milk production is 10,000 pounds per cow, some cows have attained 50,000 pounds. While pork production now averages more than four pounds of grain per pound of pork, production of a pound of pork has been achieved with an input of only two pounds of grain. A number of reproductive breakthroughs are now practically accomplished in livestock.[2] These include twinning in cattle, embryo transplants, controlled fertility periods, and sex determination, with cloning on the horizon. Conservation farming methods, particularly minimum tillage, may give a sizeable boost to yields as a simultaneous conservation measure.[3] As early as 1981, soybean varieties had been announced that produced more than 100 bushels per acre under field conditions at two locations in Iowa (20). This accomplishment may open an entirely new vista in corn and soybean yields in the Corn Belt.

Watson presented the conditions under which farmers' yield distributions for corn might be moved to the right by the year 2000.[4] His distribution has a mode at 410 bushels, with the upper tail of the distribution reaching 525 bushels. These potential yield breakthroughs, along with resource constraints in other regions imposed by groundwater depletion, are other reasons why the relative importance of agricultural production in the North Central Region should be enhanced in the future.

Many biological scientists are, given sufficient research resources and time, optimistic about the increase in yield potential from the genetic engineering of plants. The scientists propose transforming leaf structures and configurations to boost solar energy use above its current 20 percent average, gene transplants to combine various characteristics of plants, improvement of nutrient uptake, and wider tolerance to stress conditions. Some envision lifting soybean yields to those of corn under the same soil and environmental conditions and increasing protein content of corn to that of soybeans. With gene transplants from soybeans to sunflowers, some believe that biological nitrogen fixation by cereals is feasible. Perennial wheat is on the verge of becoming a reality, while peren-

[2]Private conversation, February 6, 1982, with William Hansel, Veterinary Research Tower, Cornell University.
[3]Private conversation, February 6, 1982, with Robert A. Peters, Plant Science Department, University of Connecticut.
[4]Private communication, February 24, 1982, with Andre G. Watson, Science and Technology Center, International Harvester, Inc., Hinsdale, Illinois.

nial corn already exists. In summary, there are many conflicting claims, but evidence indicates that there exists a substantial physical and biological potential for increased agricultural production.

Land Supplies

Another set of variables or conditions affecting the amount of water needed for economic agricultural production in future decades surrounds land supplies and use. Grain production in the North Central States increased 2.84 percent per year from 1937 to 1977 (22). Increased yields accounted for a large part of this production increase in all states in the region throughout the period. For the states of Illinois, Indiana, and Iowa, the contribution of land to increased grain output was greater in the first half of the period than in the second half. A continuation of this trend is expected. Thus, there is limited potential for further expansion of grain production on land presently in farms.

Further, a proposition prevails that the nation will have less land available for agriculture as this resource is transferred to nonagricultural uses and as its productivity is eroded away through exploitative farming methods encouraged by high commodity prices and land values (5). According to projections by Barlowe (3), if past trends in nonfarm uses of land continue to 2000, urban areas will use only about 2 percent of the U.S. land area. This compares to 1.5 percent in urban areas in 1969. Transportation uses would require only 1.2 percent, compared to 1.1 percent in 1969.

By 2000, only 2.7 percent or 61 million acres would be used for urban areas, highways, roads, railroads, and airports (11). This is not a mammoth transfer of land to nonagricultural uses if gauged against possibilities in zoning and a considerable supply of land that could be transferred to crops. The 1967 Conservation Needs Inventory estimated that 265 million acres could be converted to the equivalent of Class I and II cropland (35). Considering location and economic feasibility, the 1977 National Resource Inventory put this convertible supply at 127 million acres, with 40 million acres being readily convertible.[5] Hence, a net supply of land exists that could serve as a substitute for (a) further water development or (b) a decline in irrigation as water costs increase with energy prices, greater pumping depths, and competition by nonagricultural users.

The nation has become increasingly concerned with soil erosion, land productivity, and resource conservation. Passage of the 1977 Soil and Water Resources Conservation Act (RCA) is one expression of this concern. A Harris Poll, commissioned by the Soil Conservation Service, in-

---

[5]From unpublished data tapes, 1977 National Resource Inventory, Soil Conservation Service, U.S. Department of Agriculture.

dicated that 77 percent of the nation's population considered problems of soil conservation to be urgent. Hence, a question arises of whether attainment of certain soil conservation goals implied under RCA might restrict the use of marginal land and thus restrain the productivity growth through land and its substitutability for water in the future.

Iowa State University's Center for Agricultural and Rural Development recently completed an extended modeling analysis of alternative futures under RCA (9). This analysis suggests the impact on land use and productivity under high exports and various conservation goals in 2000 and 2030 and the use and imputed value of water at these times. Estimated shifts in cropping patterns are presented later.

Soil Conservation Impacts

There is growing public sentiment for greater attention to soil conservation. Since the upward burst in export demand and elimination of supply control programs, crop farming has become more extensive and in some areas approaches monoculture. This concern has been reflected in passage of RCA, national dialogue, and opinion polls. It is possible that the nation might institute a broad and more vigorous soil conservation program by 2000. Such a program could affect the absolute amount and relative shares of commodity production by agricultural regions.

In an analysis of RCA potentials, a programming model was used to project such a program to 2000 and 2030 (9). One scenario supposed that all land classes in all U.S. agricultural producing areas would have to lower erosion to soil loss tolerance (T) levels, the tolerance level at which soil formation would offset soil loss in maintaining productivity. The results of the scenario with erosion lowered to T levels were compared to a base run where no limits were placed on soil loss (Table 1). The comparison suggested that any future programs to reduce erosion to or near T levels would have little effect on the relative shares or importance of the North Central Region in crop production. As now, it would be the major region for production of feed grains, fed cattle, and hogs. Its relative importance would even likely grow as total food demand increases and as the potential of reduced water supplies constrains feed grain production and cattle feeding in the Southwest. But, as mentioned, more of the long-run increase in output from the North Central Region will likely come from new technology and the inputs representing it rather than from expanding irrigation.

Energy Price Effects

Energy price also is a variable of uncertain magnitude in the future. Generally, the real price of energy is expected to increase over the next two decades. Taken alone, higher prices for energy and inputs derived from it should dampen the use of these factors in agriculture. In the 11 major irrigating states, rising energy prices are likely to have a greater influence on pumping costs than declining groundwater levels (28). However, energy and related prices are not expected to restrain agricultural production to any greater extent in the North Central Region. There is even basis for supposing the restraint might be less severe in the North Central States because (1) a relatively small proportion of the region's agriculture is or will be irrigated and (2) the irrigated portion of the region probably has a more favorable supply-price situation for groundwater than other regions of the United States.

Whether higher real energy prices reduce the use of these inputs or have important limiting effects on productivity growth in the North Central Region (or all of U.S. agriculture) depends upon several variables and conditions. One important factor is the elasticity of production for these inputs. Use of fertilizer over the North Central Region is so high and its elasticity at these levels of use is so low that slightly reduced use would not reduce yields much. A study of the impact of a substantial increase in the relative price of energy-related inputs in the Corn Belt indi-

Table 1. Regional shares of grain production, in percent, under unrestricted and T-level soil erosion losses for the year 2000.

Zone	Base Run (unrestricted)	T Level (restrictions)
Feed Grains		
Northeast	5.1	5.0
Southeast	2.9	4.0
Lake States	22.5	20.8
Corn Belt	31.9	32.7
Delta States	2.1	1.1
Northern Plains	20.1	19.3
Southern Plains	13.7	15.1
Pacific	1.7	1.9
Wheat		
Northeast	3.8	3.4
Southeast	9.9	5.8
Lake States	10.9	9.4
Corn Belt	5.0	13.2
Delta States	13.2	7.9
Northern Plains	25.8	20.6
Southern Plains	15.5	20.8
Pacific	16.1	18.9
Oilmeals		
Northeast	0.5	0.7
Southeast	4.3	1.7
Lake States	10.5	18.2
Corn Belt	43.7	41.1
Delta States	11.4	14.3
Northern Plains	9.8	18.3
Southern Plains	19.6	5.5
Pacific	0.2	0.2

cated that corn acreage would decline less than four percent, with soybean acreage being the primary replacement (30). Similarly, there are some indications that water use levels are generally so high that production elasticities are very small—or perhaps that water marginal productivity is even negative (2, 15).

Whether higher energy prices will even lower the use of water and associated inputs over the long run and thus constrain productivity depends upon other variables and conditions. If the optimists with respect to possible future innovations and technologies are correct and if society invests enough in research and development, these technologies could interact to increase greatly the marginal productivity of energy-based inputs (i.e., markedly raise the production surface for which energy-based materials and new technologies are inputs). In a similar vein, the new technologies can respond to the relative scarcity of energy and, in a sense, serve as partial substitutes for energy-based inputs, allowing yields still to increase with relatively less energy used in the input mix. In terms of the concept of induced innovations (25), this latter set of possibilities would seem to prevail within the resource-price setting expected for the future.

The Prospects

After settlement of the nation's land at the beginning of this century, the real price of food rose. However, initiation of effective public and private research developed technologies substituting for land. These technologies, together with low real prices for water, energy, chemicals, and other capital inputs, caused the nation's agricultural supply capacity to increase greatly. Real prices for agricultural commodities, aside from during war or world drought, declined and have remained low over most of the last half century.

In the long run, we believe that growth in world food demand and resource limits and prices will cause real prices of agricultural commodities to rise. However, we believe that there will be a period of perhaps several decades when continuation of yield trends and improvement of agriculture in developing countries can maintain food supply relative to food demand. An important characteristic of this period may be volatility of prices and incomes, as experienced in the period 1978-1982. There will be dampened prices during years of generally normal weather worldwide and high prices in years of crop shortfalls in major regions of the world. Only modest increases, if any, in real prices may be realized during this period. Beyond that, however, we expect world demand to generate rising real prices of food. These rising real prices should help induce further research on potential, even exotic, technologies.

Looking far enough into the future, when water supplies to U.S. agriculture have declined considerably due to competition from other sources and decline in major aquifers to recharge rates, real prices for farm commodities may rise substantially. Some increase in irrigation in the eastern Corn Belt can then be expected, but it will be of modest magnitude because of the limited existence and capacity of aquifers and streams in this area. Increased supplemental irrigation is expected, but its magnitude relative to production and resource use in the region will be modest. More double cropping in the area could encourage supplemental irrigation to the limit of economic supplies of water.

Some further development of irrigation is expected in the western portion of the North Central Region, especially Nebraska and Kansas. This development will come as more landowners and investors attempt to capitalize on their share of available water and to maintain their asset values before greater restrictions are placed on drilling new wells and withdrawals. Eventually, though, irrigated agriculture will have to back down from the peak that will be attained in the next 20 to 30 years. Withdrawal at recharge rates will entail some shifts in agriculture, some decline in total area productivity, and extended institutional arrangements to help guarantee greater productivity and equity in water use.

Taking the North Central Region as a whole, the long-run supply of water for irrigation is not likely to be large enough to offset declines in agricultural productivity resulting from dropping water use over the entire Ogallala Aquifer area to recharge rates. Rather than from extended water use, gains in productivity in the North Central Region to meet long-run increases in domestic and export demands will need to come from technological improvements that substitute for both water and land. As mentioned, with appropriate investment in conservation practices over the long run, the North Central Region and the nation will be better able to increase land inputs than water inputs in extending output to meet future food demands. Because a declining supply of water available to agriculture is apparent, the need will prevail to use this limited supply in a manner that maintains productivity at the highest levels possible. Maintenance of productivity will require mechanisms that improve allocation among commodities, locations, and users. Obviously, the creation and implementation of such mechanisms, starting with the institutional conditions now shaping water control and allocation, will have to incorporate elements that provide equity in asset values and income of current and future water users.

In 1981 the North Central Region produced 86 percent of the nation's feed grain supply, 75 percent of soybean production, and 51 percent of wheat production. It had 75 percent of the pigs raised and 52 percent of the cattle fed in 1980. Among the important changes that could cause shifts in the shares of these commodities produced are water supplies for irrigation, soil conservation programs, and energy prices. The region's share of fed cattle declined considerably over the last two decades with the development of cattle feeding in the Southwest.

Water supplies and energy prices will augment each other in their effects on the supply price of agricultural commodities. Together, they will affect the western part of the region through increased pumping costs. However, our analyses suggest that by the year 2000 that part of the Ogallala Aquifer falling in the North Central Region will not be impacted as heavily as the southern Ogallala area (27). With an important part of national population growth taking place in the 17 western, irrigated states west and south of Kansas, Nebraska, South Dakota, and North Dakota, competition for surface water will be stronger in these western areas outside of the North Central Region. However, the North Central Region, mainly the four western states in the region, will rely more on groundwater in a relative sense than will the other 17 western states.

A normative analysis of water demand in the future suggests that a four-fold increase (over 1975 real levels) in all water costs would cause groundwater use in the four western states of the North Central Region to drop much more than the use of surface water in other states (5). However, use of groundwater in these states would drop much less than in other states of the Ogallala Aquifer area. In the Ogallala Aquifer area alone, the acreage programmed to be depleted in 2000 under continued high demands would be much larger in the area outside the North Central Region than in that area within the region. For example, the southern Ogallala area, entirely outside the North Central Region, would have 1.2 million acres depleted, while the northern Ogallala and central Ogallala areas together would have 616,000 acres depleted (27). The two more northern portions of the Ogallala area fall in the North Central Region (except for a relatively small acreage in Colorado and Oklahoma). Hence, from the standpoint of water supplies and prices, the North Central Region is expected to maintain or increase its relative share of national production because (a) the majority of the region does not, and will not in the future, use irrigation water and (b) that portion of the North Central Region using irrigation water will be less adversely affected by energy and water costs than areas outside the region.

A recent study provides additional information concerning regional shifts in crop production with substantial increases in the cost of surface and groundwater (16). The results indicate that in terms of net return from field crops the Corn Belt gains more than any other region as a consequence of such cost increases.

REFERENCES

1. Anderson, F. H, editor. 1976. *Future Report* 14(16).
2. Ayer, Harry W., and Paul G. Hoyt. 1981. *Crop water production functions; economic implications for Arizona.* Bulletin No. 242. Arizona Agricultural Experiment Station, Tucson.

3. Barlowe, Raleigh. 1975. *Demands on agricultural and forestry lands to service complementary uses.* In *Perspectives on Prime Lands.* U.S. Department of Agriculture, Washington, D.C. pp. 105-119.

4. Boyer, J. S. 1982. *Plant productivity and environment.* Science 218 (4571): 443-448.

5. Brown, Lester R. 1979. *The worldwide loss of cropland.* Worldwatch Paper 24. Worldwatch Institute, Washington, D.C.

6. Christensen, Douglas A., Andrew Morton, and Earl O. Heady. 1981. *The potential effect of increasing water prices on U.S. agriculture.* CARD Report No. 101. Center for Agricultural and Rural Development, Iowa State University, Ames.

7. Crosson, Pierre. 1979. *Agricultural land use: A technological and energy perspective.* In Max Schnepf [editor] *Farmland, Food and the Future.* Soil Conservation Society of America, Ankeny, Iowa.

8. Crosson, Pierre R., editor. 1982. *Cropland crisis: Myth or reality?* Johns Hopkins University Press, Baltimore, Maryland.

9. English, Burton C., and Earl O. Heady. 1980. *Short and long-term analysis of the impacts of several soil loss control measures in agriculture.* CARD Report No. 93. Center for Agricultural and Rural Development, Iowa State University, Ames.

10. Food and Agricultural Organization of the United Nations. 1982. *Agriculture, 1981. Towards 2000.* Rome, Italy. pp. 6-64.

11. Frey, H. T. 1979. *Major uses of land in the United States: 1974.* Agricultural Economic Report No. 440. Economics, Statistics, and Cooperatives Service, U.S. Department of Agriculture, Washington, D.C.

12. Fuller, R. 1980. *More and more effort yields less and less.* Des Moines Register and Tribune (February 17).

13. Heady, Earl O. 1980. *Economic and social conditions relating to agriculture and its structure to 2000.* Organization for Economic Cooperation and Development, Paris, France.

14. Heady, Earl O. 1982. *The adequacy of agricultural land: A demand-supply perspective.* In Pierre R. Crosson [editor] *The Cropland Crisis: Myth or Reality?* Johns Hopkins University Press, Baltimore, Maryland. pp. 23-56.

15. Hexem, Roger, and Earl O. Heady. 1976. *Water production functions in irrigated agriculture.* Iowa State University Press, Ames.

16. Lacewell, Ronald D., and Glenn S. Collins. 1982. *Implications and management alternatives for western irrigated agriculture.* Technical Article 17807. Texas Agricultural Experiment Station, College Station.

17. Lee, John. 1983. *Some consequences of the new reality in U.S. agriculture.* In David E. Brewster, Wayne D. Rasmussen, and Garth Youngberg [editors] *Farms in Transition.* Iowa State University Press, Ames. pp. 3-22.

18. Lin, K. T., and S. D. Seaver. 1978. *Were crop yields random in recent years?* Southern Journal of Agricultural Economics 10(2): 139-142.

19. Lu, Yao-chi, Philip Cline, and Leroy Quance. 1979. *Prospects for productivity growth in U.S. agriculture.* Agricultural Economic Report No. 435. Economics, Statistics, and Cooperatives Service, U.S. Department of Agriculture, Washington, D.C.

20. Muhm, Don. 1982. *100-bushel beans grown in Iowa tests.* Des Moines Register (February 22): 13.

21. National Academy of Sciences. 1975. *Agricultural production efficiency.* National Research Council, Committee on Agricultural Production Efficiency, Board on Agriculture and Renewable Resources, Washington, D.C.

22. Patrick, George F., and Earl R. Swanson. 1979. *Components of growth in grain production in the North Central States: 1937 to 1977.* North Central Journal of Agricultural Economics 1(2): 87-96.

23. Pfeifer, R. P. 1976. *Record yields and your operation.* Crops and Soils 28: 5-7.

24. Ruttan, Vernon W. 1979. *Inflation and productivity.* American Journal of Agricultural Economics 61(5): 896-902.

25. Ruttan, Vernon W. 1982. *Agricultural research policy.* University of Minnesota Press, Minneapolis.

26. Schnittker, John A. 1981. *Future of the United States Department of Agriculture.* In Proceedings, Sixth Annual Midwestern Conference on Food, Agriculture, and Public Policy. Business Government Interaction, Sioux City, Iowa.

27. Short, Cameron, Anthony F. Turhollow, Jr., and Earl O. Heady. 1981. *Regional impacts of groundwater mining from the Ogallala Aquifer with increasing energy prices, 1990 and 2000.* CARD Report No. 98. Center for Agricultural and Rural Development, Iowa State University, Ames.

28. Sloggett, Gordon. 1981. *Prospects for ground-water irrigation: Declining levels and rising energy costs.* Agricultural Economics Report No. 478. Economic Research Service, U.S. Department of Agriculture, Washington, D.C.

29. Sundquist, W. Burt, Kenneth M. Menz, and Catherine F. Neumeyer. 1982. *A technology assessment of commercial corn production in the United States.* Special Bulletin 546. Agricultural Experiment Station, University of Minnesota, St. Paul.

30. Swanson, E. R., and C. R. Taylor. 1977. *Potential impact of increased energy costs on the location of crop production in the Corn Belt.* Journal of Soil and Water Conservation 32(3): 126-129.

31. Swanson, E. R., and J. C. Nyankori. 1979. *Influence of weather and technology on corn and soybean yield trends.* Agricultural Meteorology 20: 327-342.

32. Swanson, E. R., and J. C. Nyankori. 1981. *Influence of weather and technology on corn and soybean yield trends—reply.* Agricultural Meteorology 23: 175-180.

33. Swanson, Earl R. 1981. *Agricultural productivity and technical progress: Acceleration or atrophy?* Staff Paper No. 81 E-152. Department of Agricultural Economics, University of Illinois, Urbana.

34. Tammeus, W. D. 1975. *Evidence of growth limits in agriculture seen.* The Kansas City Star (November 2).

35. U.S. Department of Agriculture. 1971. *National inventory of soil and water needs, 1967.* Statistical Bulletin 466. Washington, D.C.

36. Walker, W. M., E. R. Swanson, and S. G. Carmer. 1982. *Soybean yields continue to increase.* Better Crops 66(Spring): 20-21.

37. Wittwer, Sylvan H. 1980. *Agriculture in the 21st century.* In Proceedings, Agricultural Sector Symposia. World Bank, Washington, D.C.

38. Wittwer, Sylvan. 1982. *New technology, agricultural production and conservation.* In Harold Halcrow, Earl O. Heady, and Melvin Cotner [editors] *Soil Conservation, Policies, Institutions, and Incentives.* Soil Conservation Society of America, Ankeny, Iowa.

6

Irrigation Management: Current and Prospective Issues

Arlo Biere and Frederick Worman

To consider the current and prospective issues associated with irrigation management in the North Central Region, we shall divide the region into two subregions. The humid subregion includes the Corn Belt and Lake States. The semiarid subregion encompasses the High Plains States. Nearly 85 percent of the irrigated acres in the North Central Region are in the Plains States, and more than 82 percent of the irrigated acres in the North Central Region are in Nebraska and Kansas. In fact, over half the irrigated acres in the region are located in Nebraska alone.

Irrigation in the humid subregion is supplemental. Natural precipitation provides the bulk of the soil moisture used by growing crops. Irrigation is used because rainfall is not adequate or reliable during the crop's critical growth stage; because the coarse texture of the soil in the root zone offers a low soil moisture-holding capacity, which necessitates frequent waterings to maintain crop growth; or because the crop grown in an area may be sensitive to soil moisture stresses. Corn or soybeans are grown on more than half the acres irrigated in the humid region. The remaining acres are in potatoes, tobacco, and fruits and vegetables.

Full irrigation is typical in the semiarid Plains States. Natural precipitation generally is insufficient to support the crop even in years of average rainfall. The predominant crop grown under irrigation is corn. Until irrigation was introduced, corn was not an economically viable crop in the semiarid areas of western Kansas and Nebraska. Other irrigated crops grown in those states include grain sorghum, wheat, soybeans, pasture, and hay.

Most irrigation water is obtained by pumping from aquifers that have a recharge less than the annual withdrawal for irrigation and other uses. That has led to concern about the eventual depletion of groundwater supplies. The major aquifer supplying irrigation water in this area is the Ogallala Aquifer, which extends from Nebraska south into Colorado, Kansas, Oklahoma, New Mexico, and Texas. Of these states, only Nebraska and Kansas are part of the North Central Region.

The purpose of irrigation obviously differs in the two subregions; consequently, irrigation management issues differ. In the humid region, irrigation is used mainly to improve the distribution of water reaching the root zone. Management issues concern the proper timing of irrigations given uncertain weather conditions and the impacts of irrigation on stream and groundwater quality.

In the semiarid subregion, irrigation provides soil moisture to grow crops that naturally require more humid conditions. The primary concern is depletion of groundwater supplies. In some locales the issue of groundwater and stream pollution through nitrogen leaching from the root zone to the aquifer is critical also.

Irrigation Management Objectives

What are the current and prospective irrigation management issues in the region, and how are these objectives being dealt with?

First is the issue of irrigator objectives. These objectives are most important because they relate to the individuals who are actually irrigating.

Probably next in importance are the objectives of the business community that depends upon irrigated agriculture. This second group might also include irrigation district associations, federal and state agencies, and university or college programs, the existence of which depends upon irrigation. These are included in the second group because of the general tendency for all in this group to see that the positive benefits of irrigation outweigh the negative ones.

A third group encompasses those citizen groups concerned for the long-term impacts of irrigation on the environment, people, and society generally. This group may include individuals in government and academia. But, those from academia are more likely not in colleges of agriculture.

While we use this classification for discussion purposes, we do not deny that one group does not hold an objective held by another group. Rather, it is the weights that each group attaches to various objectives that differ. For example, it would be hard to believe that farmers in Hall County, Nebraska, are not concerned about groundwater quality when the water from wells on many farms is no longer fit for human consumption, at least for infants, because of high nitrate levels.

The individual irrigator is the primary decision-maker influencing irrigation development, Much can be learned about current and prospective issues in irrigation by looking at the objectives of these individuals, but studies of the objectives of the individual irrigator are not available. One objective of an irrigator, of course, is to maximize profit. But other objectives may have considerable influence. Risk aversion and how it affects decision-making is one. Another may involve the use of family labor only. Still other preferences may influence irrigators' decisions, for example, maximizing yield per acre.

To date, analyses of irrigator's decisions have not adequately looked at irrigators' objectives. For example, studies of irrigation investment decisions consider only economic return from investment. Other analyses concern irrigation scheduling. Irrigations should be scheduled to maximize or best satisfy the objectives of the irrigator. Because of the complexities in calculating schedules, most irrigation scheduling is based only on the soil moisture depletion criterion. How well does that criterion produce results that satisfy the irrigator's objectives? That question can be answered as models are built that relate scheduling to those objectives.

The objectives of the other two groups in the decision-making chain can be classified as promoting regional economic growth or maintenance of the economic base through irrigation and avoiding adverse environmental impacts from irrigation, respectively. Conflicts among the objectives are resolved through the political and judicial processes.

In the case of each group, it is important for economists to identify the implicit objectives by which decisions are being made and to develop models that can explicitly explore the impact of those objectives on decision-making. The result will be a better understanding of the decision-making process and the proposal of better management models.

Irrigation Management Factors

Six factors have impacts on irrigation management: declining groundwater and surface water supplies; the rising real cost of energy; scheduling problems, with uncertainty regarding the relation between water applied and crop yield; technical developments in instrumentation and microcomputers; adverse environmental impacts from irrigation; and supplemental irrigation. These factors affect irrigation management today and will have substantial impacts in the future. The issue of adverse environmental impacts, in the form of nonpoint pollution, may become more widespread with more intensive irrigated crop production on light soils.

Declining Groundwater Supplies

The major irrigation management factor in the High Plains is declining water availability because of declining groundwater tables in the

Ogallala Aquifer and declining irrigation water yields from surface sources. In humid areas, the potential problem is competition with other water users. Except for small areas in the humid region, the question of water availability is confined mainly to the semiarid Plains.

Congress, in Public Law 94-587, authorized a study of decreasing water supplies and the probable impacts of such declines on local economies in the High Plains. The federal study involved a six-state region. State studies were conducted by universities and state agencies. These studies were coordinated by a general contractor. The general contractor also developed a regional analysis for the entire study area.

Alternative development strategies included voluntary water demand management encouraged through promotion of water conservation strategies and application of proven water-saving technologies; mandatory water demand management; water supply augmentation—local, intrastate, and interstate; and alternative nonagricultural development and use of resources. Except in certain local situations, water supply augmentation is not economically feasible. The farther the water must be transported, the more uneconomical this alternative is for High Plains' agriculture. For example, the Nebraska study of in-state, interbasin water transfers (32) showed that the cost of delivering such water would be two to seven times more than irrigators could afford to pay. (Estimated annual cost per acre-foot of water ranged from $189 to $750, depending upon the transfer project.)

In both Kansas and Nebraska, investigators evaluated the alternatives of reduced water use through voluntary efforts to improve irrigation efficiency and mandatory water demand reductions. In Kansas, increasing irrigation efficiency resulted in a higher production value in each year than would occur under current conditions or under water demand management (24). The Nebraska study showed that increasing irrigation efficiency through incentives would produce the highest return to land and management during each year from 1977 to 2020. But that strategy does little to conserve groundwater. According to the Nebraska report (32), the improved efficiencies used in that strategy affected net withdrawals only slightly even though gross pumpage was reduced considerably. The reason: With higher efficiencies there would be less excess water returned to the aquifers by deep percolation.

The other alternative considered in both states was reduction of water use through mandatory water management. The Kansas results (24) projected a reduction in the value of production, compared with the previous alternative. The value of production was below that for the baseline in the early years and above that for the baseline in later years. The Nebraska results were somewhat different. For each of the years from 1977 to 2020, the results showed a reduction in net returns to land and management from that projected under baseline conditions. We suspect two factors led to the different results in the two states. First, Nebraska, in general, has a larger groundwater supply and will not exhaust the groundwater as soon as Kansas. Second, the Nebraska study assumed a faster rate of increase in irrigation efficiency than the Kansas study. Prior to 2020, the Nebraska study results do not show economic returns to demand management.

We drew two major conclusions from the study: First, the most economically viable alternative is to make the best use of the available groundwater, i.e., do not control groundwater use. Second, those states that will exhaust their groundwater resources should plan to return to dryland farming.

Rising Energy Prices

Rising energy prices affect irrigation management, particularly when the total lift or head required to deliver the water to the field is high. We believe that all of the economic information required to consider the impacts of rising energy prices is contained in the ratio of the price of energy to the price of crop harvested from the irrigated acres. The rule for optimal use of a resource is to use the resource to that intensity where the value of the marginal physical product of the resource equals the unit price of the resource. But the value of the marginal physical product is given by the product of the marginal physical product of the resource and the price of the commodity produced. The rule for optimal resource use can thus be restated as follows: Use the resource to the intensity such that the marginal physical product of the resource equals the ratio of the price of the resource to the price of the commodity produced. As long as the technology does not change, the intensity of resource use is inversely related to the ratio of the price of the resource to the price of the product. That ignores the impact of relative price changes of other resources used.

Our argument is that the impact of rising energy prices can be found by looking at the ratio of the price of energy to the price of the product. Tables 1 and 2 show the actual and projected prices for four energy sources expressed as a ratio of the price of corn for the same year. Ratios through 1981 are of actual average prices. Ratios after 1981 are based on projected prices used in the High Plains study. The differences between the 1977 ratios in table 1 and table 2 are the result of different sources of information used for the two tables. Projected corn prices for the High Plains study were provided by the U.S. Department of Agriculture's Economic Research Service through the National Interregional Agricultural Projections Model. The energy price projections were made by the general contractor for the High Plains study.

We used the price of corn rather than the price of another irrigated crop because, of the crops considered in the High Plains study, corn was projected to show the greatest real price increase between 1977 and 2020. Over that period, the real price of corn is projected to increase 9 percent more than the real price of sorghum and nearly 40 percent more than the real price of alfalfa, hay, or wheat. During the same period, the real price of soybeans is projected to decline more than 25 percent from the 1977 real price.

Tables 1 and 2 confirm that energy prices are now rising faster than corn prices and will continue to do so in the future. In fact, rising energy prices may have nearly as profound an impact on High Plains irrigation as the decline in groundwater. Scheduling studies indicate that the optimal irrigation schedule for a given crop is only moderately responsive to changes in the variable cost of irrigating. Corn growers may be forced to switch to crops that use water less intensively for two reasons. First, rising irrigation costs and high water requirements for corn will make corn less profitable than other crops for that irrigator. Second, it will not be possible to reduce water use on corn sufficiently to adjust for the higher irrigation costs. In some cases, farmers may be forced to return to dryland production because of the high cost of irrigating or because of the eventual lack of groundwater. The real price of energy will continue to rise faster than the real price of the crops harvested from the irrigated acres (Table 2).

Crop Response Uncertainty

Many factors influence the relationship between applied water and yield. In fact, we believe that there are so many factors involved that any

Table 1. Actual energy expressed as a ratio of the price of corn during the same year (42, 43).

Year	Diesel Fuel/Corn	Electricity/Corn	Propane/Corn	Natural Gas/Corn
1965	.142	.020	.11	NC*
1970	.136	.016	.11	NC
1977	.222	.018	.19	NC
1981	.468	.022	.28	NC

*Not calculated due to lack of consistent price information comparable to data in table 2.

Table 2. Projected energy price expressed as a ratio of the price of corn during the same year (32).

Year	Diesel Fuel/Corn	Electricity/Corn	Propane/Corn	Natural Gas/Corn
1977	.200	.015	.16	.68
1985	.353	.018	.20	1.39
1990	.352	.019	.25	2.02
2000	.341	.020	.24	1.96
2020	.343	.021	.25	2.01

relationship between applied water and yield provides little useful management help (*14*). Obvious factors include the crop's physiological characteristics, climatic conditions, cultural practices, heterogeneity of a field's soil, and heterogeneity of soil moisture levels throughout the field due to differences in soil topology, soil texture, and moisture received. Weather conditions are unpredictable. Crop response to climate and soil moisture is not entirely predictable. Also, there are uncertainties due to within-field variations.

Of what importance is crop response uncertainty to irrigation management? Such uncertainties, we believe, create tendencies for irrigators to overirrigate. Irrigation scheduling models must account for that uncertainty to provide reliable management information. Finally, such uncertainties mean that we must develop more accurate crop models.

A common point made by scientists and irrigaton specialists is that a farmer's tendency to overirrigate may be because the farmer is substituting a cheap input, water, for some more scarce and expensive input, such as labor or capital. Beyond that, however, uncertainty may be the reason for overirrigating. One case study of such uncertainty found that a risk-averse farmer could be expected to use more water per acre and put less land into production if water availability were limited (*14*).

Consider the rule for optimal use of a resource. That rule is to use the resource to the level of intensity where the value of the marginal physical

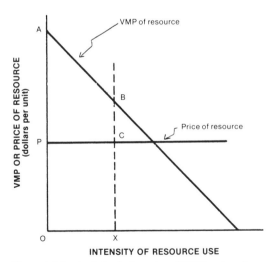

Figure 1. Marginal physical product (VMP), price of resource, total value of resource used in production, cost of resource used, and net value of resource used.

product of the last unit applied equals the price of the resource. Based on
the notion from marginal productivity theory that the marginal physical
product curve becomes the demand curve for the resource, the area under the curve from the origin to the level of use of the resource gives the
value of the use of that amount of resource (Figure 1, the area bounded
by OABX). The cost of the resource used is given by the area under the
price line from the origin to the level of use of the resource (Figure 1, the
area bounded by OPCX). Net return to this use of that resource is then
given by the difference between the two areas (Figure 1, the area bounded by PABC). Maximum net return from the use of this resource is obtained at the point where the marginal physical product equals the price
of the resource shown (Figure 2, intensity level Z).

The shaded area between the marginal physical product curve and the
price line can be used to find the net loss associated with a resource use
level different from the optimal level of resource uses. For example, for a
given resource use of Z − a (Figure 2), the loss associated with nonoptimal use of the resource is given by the shaded area between resource use
Z − a and Z. A linear marginal physical product curve gives a symmetrical loss function. For example, with the linear curve shown in figure 2,
the net loss associated with using Z + a is the same as that associated with
using Z − a. For a producer to overuse or underuse this resource by the
same quantity produces the same loss compared with the return from using the optimal amount.

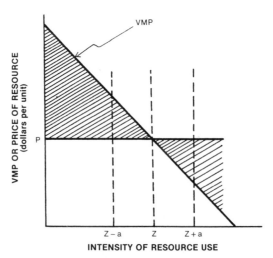

Figure 2. Loss associated with nonoptimal resource use
[marginal physical product (VMP) linear].

To have a symmetrical loss function, the marginal physical product curve must be linear, which means the production function must be a quadratic. Because the curve is a multiple of the first derivative of the production function with respect to the resource input, the linear marginal physical product curve implies that the production function must be a quadratic function of that resource input. Quadratic production functions have been estimated for agronomic production functions where the water input variable is defined either as total seasonal water use or as total water applied (22). With such functions the loss function is symmetrical.

For the irrigation scheduling decisions facing the irrigator, it is unreasonable to assume that the production function is a quadratic function. When producing a crop under irrigation, irrigation water cannot be viewed as a homogeneous input. That is not to argue that there are differences in the physical quality of the water applied to a field. Rather, the time when the water is applied influences yield. Different distributions of the same total depth of water during the growing season will produce different yields. Thus, water applied at different times should be treated as different inputs to the production process.

Crop development is a cumulative process. Growth in a given period depends upon the current condition of the crop, which in turn depends upon previous growing conditions, including soil moisture conditions. Consequently, the influence of each irrigation depends upon other irrigations, even though each irrigation is, in effect, a different input to the production process. A multiplicitive production function will meet these requirements.

There still is the question of what that functional form might be. A simple functional form might be of the exponential type:

$$Y = A \prod_{t=1}^{T} X_t^{\beta_t}$$

where Y is the harvestable yield, X_t is the amount of water applied in time period t, and β_t and A are coefficients. Then, the marginal physical product curve for X_t would be

$$VMP = P \frac{\partial Y}{\partial X_t} = P \frac{A\beta_t}{X_t} \prod_{t=1}^{T} X_t^{\beta_t}$$

where P is the unit price of the harvested crop. Here the curve is not linear.

If $\beta_t < 1$, then

$$\frac{\partial VMP}{\partial X_t} < O \text{ and } \frac{\partial^2 VMP}{\partial X_t^2} > O.$$

Such a marginal physical product curve would appear as in figure 3, and the loss function is not symmetrical. The loss associated with under-irrigating is greater than the loss associated with overirrigating. Such a conclusion is not inconsistent with the evidence. Irrigation specialists have observed a tendency for farmers to overirrigate. Also, irrigators say that they would prefer to err on the side of applying too much rather than too little water.

The implications of such conditions are twofold. First, irrigation scheduling models are needed to account for such uncertainty and producer response to that uncertainty. One method dealing with risk and uncertainty is stochastic dominance theory (21) used to study irrigation scheduling. Second, models that reduce the amount of uncertainty regarding crop response are needed. Such work is needed to develop irrigation technology that will meet projected irrigation efficiencies. In the Kansas High Plains study, irrigation efficiency with flood irrigation in 1977 was estimated to be 60 percent. In that study, the researchers assumed that irrigation efficiency in 2020 for flood irrigation will have improved to 77.2 percent under baseline conditions and to 92 percent with a voluntary strategy to increase irrigation efficiency (24). Irrigation efficiencies for sprinkler irrigation in 2020 are projected to be higher than those for flood irrigation.

Assumptions in the Nebraska High Plains study anticipate improvements in irrigation efficiency to come about slightly faster than in Kansas. Such projections of irrigation scheduling improvement will require a

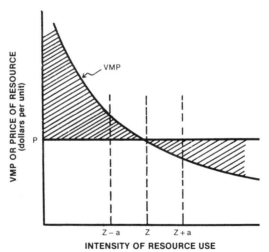

Figure 3. Loss associated with nonoptimal resource use [marginal physical product (VMP) nonlinear].

reduction in uncertainty of crop response and management strategies to deal with the remaining uncertainty.

Technological Developments

We anticipate there will be technological developments in soil moisture management, irrigation techniques, crop varieties, and equipment to monitor and evaluate cropping conditions and to determine when to irrigate for optimum results. The first three developments will take place in agronomy and agricultural engineering. The last development has an impact on engineers, agronomists, and economists working together on instrumentation, data transmission, and data processing. With these developments will come new opportunities to develop management models for use by irrigators. More reliable and less expensive instrumentation will make it possible for irrigators to monitor growing conditions in the field more carefully. The increasing availability and power of microcomputers will increase the use of more complex models on the farm. Those studying irrigation should remain alert to these possibilities. Because irrigation management has many interdisciplinary considerations, scientists working on irrigation management must be willing to join with scientists in other disciplines to exploit the potentials offered by the technological developments that are originating, for the most part, outside of agriculture.

Adverse Environmental Impacts

Irrigation can have adverse impacts on the environment through leaching of chemicals into the groundwater, through increased wind and water erosion, through exhausting groundwater supplies more quickly and, by destroying a unique fragile environment, such as the Nebraska's Sandhills.

When irrigation occurs on light, sandy soils, the potential for chemicals leaching into the aquifer can be high, especially in the case of nitrogen. Eventually, leaching leads to high nitrogen levels in the groundwater. In Hall County, Nebraska, north of the Platte River, for example, nitrogen concentrations in some wells are so high that the water is unfit for human consumption, especially infants. The Cooperative Extension Service at the University of Nebraska, in cooperation with the Agricultural Stabilization and Conservation Service, the Soil Conservation Service, and the Central Platte Natural Resource District, is working with farm cooperators in the area on a demonstration project. Scientists are calculating the nitrogen to be applied and the date and amount of water to be applied to meet a yield goal set by the cooperator, while attempting to minimize nitrogen infiltration into the aquifer (9).

Development of center pivot irrigation in Nebraska's Sandhills presents several environmental issues. These issues concern the loss of aesthetic values associated with the native range, the potential for erosion on the unprotected lands when tilled, and the potential leaching of agricultural chemicals into the aquifer. Another issue that stems from the development is that of corporate ownership of farmland. Nebraska voters recently approved an initiative to prohibit such participation by large corporations.

Supplemental Irrigation

Irrigation in the more humid areas of the North Central Region is mainly supplemental. Irrigation specialists in the more humid states confirmed (in telephone interviews) that irrigation tended to be of the supplemental type on light soils, using center pivot irrigation systems. Expansion of irrigation in these states has slowed greatly because of low crop prices.

Supplemental irrigation presents several management questions. First, is it economical? A recent study of supplemental irrigation in the subhumid area of southeastern Kansas indicated dryland grain sorghum is more profitable, given current crop prices (47). The system considered application of reservoir water with a center pivot system on heavy soils. Such a system might become profitable sooner on light soils.

Supplemental irrigation in the Corn Belt Region tends to be on light soils, where nonpoint pollution may occur easily. Management techniques are needed in these areas to avoid pollution of aquifers, streams, and lakes.

Still another consideration is timing of irrigations. An irrigation just before a rain may reduce the effectiveness of the natural precipitation. On the other hand, waiting too long to irrigate, while expecting rain, may reduce the effectiveness of the irrigation. Such questions of risk and uncertainty regarding precipitation can be studied using stochastic dynamic programming.

Modeling for Irrigation Management

Irrigated crop production involves elements of agronomy, engineering, and economics. Pest management problems should be considered also, although, until recently, these problems have been treated independent of the other aspects of irrigated crop production. Although management of irrigated crop production is a multidisciplinary problem, scientists in each field have developed models that address those aspects of most interest to their discipline. Agronomists have concentrated on the soil-water-plant relationships and plant physiological processes. Agricultural economists have concentrated on management decision models under the assumption that the biological and physical relationships are known. Agricultural engineers have concentrated on the development and operation of irrigation delivery systems and on crop response to irrigation. But the nature of the problem calls for more integrative work. Then, the full benefits of those scientific efforts can be realized by the irrigator and by policymakers who deal with issues affecting irrigation, such as the future of irrigated agriculture in the High Plains.

By far the most successful application of irrigation models has been the use of climatic data in evapotranspiration models to estimate available soil moisture throughout the growing season. Yet many opportunities to improve irrigation management by using crop response and multicrop management models are emerging.

Model development would be simplified greatly if there were a straightforward input-output model relating applied water to yield. Unfortunately, that is not possible because of the continuous way in which water availability affects a plant's growth and development. A review of agronomic studies of factors influencing yield concluded that (1) yield reductions can occur because of either depleted soil moisture or severe atmospheric evaporative demands, (2) each crop has a unique set of stages of plant development with regard to soil-water stress that must be considered, and (3) the daily effects of soil-water and atmospheric stress vary from stage to stage for any crop and differ from crop to crop (26). Such conditions call for dynamic models of plant growth that must deal with the complex way that water affects growth.

An irrigator's choice of objective function raises further questions. As discussed, the irrigator may hold one of several possible objectives. For some, the objective may be maximum profits, for others, maximum yield or minimum risk. In subsistence agriculture in third-world countries, the objective may be to minimize the risk of starvation due to crop failure. Each objective presents a different management problem. For example, the goal to maximize profit requires knowledge of both the cost of applying each increment of water at any time in the growing season and of the additional revenue resulting from the increased yield associated with each increment of water applied on any day in the growing season (6). However, if water is scarce and crop production has high priority, the objective may be maximum water use efficiency, i.e., maximum production per unit of water.

There are two major approaches to crop response modeling: (1) to estimate the crop response function or functions statistically from experimental data and (2) to build from known physiological and chemical relationships a model that describes the operation of the crop and from which the crop's response can be obtained. Each method has its limitations. Statistical models tend to ignore fundamental physiological characteristics that would influence the selection of the form of the model used. Physiological models are difficult to use in optimization models.

One discussion of water production functions (7) concluded that while water usually is considered the independent variable another variable, time, is important also. The time of application of the water input can shift the ideal production function to any of its three (economic) stages. Supplying water to corn early in the silking stage could result in increasing returns, while irrigating when the water table is high might result in negative returns.

Early production function studies of water use are susceptible to this time problem. Dividing the growing season into periods and estimating water coefficients for each period (8) overcomes the time dimension.

Agronomists and engineers estimate crop response to water based on annual water use. Often a critical or stress-day component is included to derive crop coefficients for use in generalizing the model. There are many examples of these models (5, 13, 18, 23, 29, 31, 37, 38, 39, 40).

Simulation models have two general orientations, those based on synthetic functions and data (1, 2, 3, 7, 26, 28, 34) and those that simulate actual plant growth, based on experimental data (10, 12, 17, 25, 33, 41).

Other dynamic models have been estimated from experimental data. One, a grain sorghum growth model, is a multiequation physiological model (4). Another corn growth model was developed from experimental data using spline regression (30).

Many growth models do not have economic components. Some use maximization of yield as the optimization criterion. Such models are static and give no information on irrigation scheduling. They indicate at which growth stages irrigation is necessary to keep from losing yield.

Several of the models listed above, or models incorporating these growth models, have specific economic components. The models use various optimization procedures to provide control over irrigation timing and amount of water applied. Most provide an optimum irrigation schedule based on maximization of net economic return, although several could be modified to provide optimization using other criteria, such as maximizing production for a given amount of water.

One of the major optimization techniques found in irrigation work is dynamic programming. The deterministic dynamic programming model can allocate limited quantities of water to maximize profits when there are delivery system constraints. However, the method is of limited usefulness because of computational problems when there is more than one state variable or more than one decision variable. Several researchers have used dynamic programming in the single crop situation (6, 7, 8). An alterative is optimal control theory (19, 20), including a closed loop control system (1) and an open loop stochastic control system (46), to schedule irrigations.

Modifications to single crop models show potential for increasing the usefulness of these models. Three studies have dealt specifically with weather. One included weather as a stochastic variable using a Monte Carlo technique (1). Another used a precipitation planning policy component based on historical precipitation probabilities (7). The third model incorporated rainfall predictions using five-day forecasts (31).

Other research has looked at the impact of salinity on crop systems and included this factor in crop models (11, 45). One model optimized irrigation with saline water (45). In another case, nitrogen was introduced into a corn model (44).

Three studies have dealt with uncertainty, one through a case study (14), another through estimates of expectation-variance frontiers (7), and a third through the use of stochastic dominance theory (19, 21).

Multicrop models are useful for planning cropping patterns, distributing limited water supplies, planning irrigation projects, analyzing various possible water use regulations in terms of actual effect on production, and analyzing interregional water use efficiency. Generally, the decisions to be made include (1) which crops should be planted, (2) what amount of the available land should be dedicated to each crop, and (3) how much water should be allocated to each crop. The inputs for these models are some type of single crop model and various constraints, such as the total amount of water available and when it is available, total land available, and maximum or minimum amounts of each crop.

Several simulation models have been used in the multicropping context. These include simulation models to determine allocation of water (3, 26, 36). One simulation model was used in conjunction with a river basin model to determine yields and estimate net return per unit area for each crop and for an entire subbasin (15). Linear programming has been used in the research efforts to determine optimal allocation of water between crops (7, 16, 27), and a Kansas State University study looked at how water might be allocated among crops as the yield of an irrigation well declines (35). That study found that with a well yielding 1,200 gallons of water per minute a farmer would put most of a 160-acre field into irrigated corn production with some irrigated wheat and grain sorghum. As the yield of the well dropped to 200 gallons per minute, the farmer raised fewer acres of irrigated corn and more acres of irrigated sorghum and irrigated wheat. Wheat and sorghum were used in the solution because the two crops would not require irrigation during the same period.

A Final Thought

Climatic conditions in the North Central Region range from humid to semiarid. Irrigation management issues vary between the two types of climatic conditions.

Much research remains to be done in irrigation management. Because of the broad range of factors that need to be considered in irrigation management, cooperative work among scientists from many different disciplines is needed to provide the answers to various management issues.

REFERENCES

1. Ahmed, J., C.H.M. van Bavel, and E. A. Hiler. 1976. *Optimization of crop irrigation strategy under a stochastic weather regime: A simulation study.* Water Resources Research 12: 1,241-1,247.
2. Anderson, R. L. 1968. *A simulation program to establish optimum crop patterns on irrigated farms based on preseason estimates of water supply.* American Journal of Agricultural Economics 50: 1,586-1,590.
3. Anderson, R. L., and A. Maass. 1971. *A simulation of irrigation systems: The effects of water supply and operating rules on production and income on irrigated farms.* Technical Bulletin 1431. Economic Research Service, U.S. Department of Agriculture, Washington, D.C.
4. Arkin, G. F., R. L. Vanderlip, and J. T. Ritchie. 1976. *A dynamic grain sorghum growth model.* Transactions, American Society of Agricultural Engineers 19(4): 622-626, 630.
5. Bielorai, H., and D. Yaron. 1978. *Methodology and empirical estimates of the response function of sorghum to irrigation and soil moisture.* Water Resources Bulletin 14(4): 966-977.
6. Biere, A. W., E. T. Kanemasu, and S. Lee. 1981. *Modeling crop response for economic water use and for water conservation.* Report Number 220. Kansas Water Resources Research Institute, Manhattan.
7. Blank, H. G. 1975. *Optimal irrigation decisions with limited water.* Ph.D. dissertation. Colorado State University, Fort Collins.
8. Burt, O. R., and M. S. Stauber. 1971. *Economic analysis of irrigation in subhumid climate.* American Journal of Agricultural Economics 53: 33-46.
9. Buttermore, G., D. Eisenhauer, C. Bourg, K. Frank, and others. 1982. *Hall County water quality special project 1979, 1980, and 1981 report.* Cooperative Extension Service, University of Nebraska, Lincoln.
10. Childs, S. W., J. R. Gilley, and W. E. Splinter. 1976. *A simplified model of corn growth under moisture stress.* Paper 76-2528. American Society of Agricultural Engineers, St. Joseph Michigan.
11. Childs, S. W., and R. J. Hanks. 1975. *Model of soil salinity effects on crop growth.* Soil Science Society of America Proceedings 39(4): 617-622.
12. Curry, R. B. 1971. *Dynamic simulation of plant growth: Part I, Development of a model.* Transactions, American Society of Agricultural Engineers 14: 945-949.
13. Doorenbos, J., and H. H. Kaasam. 1979. *Yield response to water.* Irrigation and Drainage Paper 33. Food and Agriculture Organization, United Nations, Rome, Italy.
14. English, M. J. 1979. *Use of crop models in irrigation optimization.* Paper 79-4521. American Society of Agricultural Engineers, St. Joseph, Michigan.
15. Fapohunda, H. O., and R. W. Hill. 1981. *River basin hydro-salinity-economic modeling.* Journal of the Irrigation and Drainage Division, American Society of Civil Engineers 107(1): 53-69.
16. Flinn, J. C. 1971. *The simulation of crop-irrigation systems.* In J. B. Dent and J. R. Anderson [editors] *Systems Analysis in Agricultural Management.* John Wiley & Sons Australasia Pty Ltd, Sydney.
17. Hanks, R. J. 1974. *Model for predicting plant yield as influenced by water use.* Agronomy Journal 66: 660-665.
18. Hanks, R. J., H. R. Gardner, and R. L. Florian. 1969. *Plant growth-evapotranspira-*

tion relations for several crops in the Central Great Plains. Agronomy Journal 61: 30-34.

19. Harris, T. R. 1981. *Analysis of irrigation scheduling for grain sorghum in the Oklahoma Panhandle.* Ph.D. thesis. Oklahoma State University, Stillwater.

20. Harris, T. R., and H. P. Mapp. 1980. *A control theory approach to optimal irrigation scheduling in the Oklahoma Panhandle.* Southern Journal of Agricultural Economics 12: 165-171.

21. Harris, T. R., and H. P. Mapp, Jr. 1980. *Irrigation scheduling in the Oklahoma Panhandle using stochastic dominance theory.* American Agricultural Economics Association, Urbana, Illinois.

22. Hexem, R., and E. Heady. 1978. *Water production functions for irrigated agriculture.* Iowa State University Press, Ames.

23. Jenson, M. E., J. L. Wright, and B. J. Pratt. 1971. *Estimating soil moisture depletion from climate, crop, and soil data.* Transactions, American Society of Agricultural Engineers 14(5): 954-959.

24. Kansas Water Office. 1982. *Ogallala Aquifer study in Kansas: Summary.* Topeka.

25. Mapp, H. P., and V. R. Eidman. 1976. *A bioeconomic simulation analysis of regulating groundwater irrigation.* American Journal of Agricultural Economics 58(3): 391-402.

26. Mapp, H. P., V. R. Eidman, J. F. Stone, and J. M. Davidson. 1975. *Simulating soil water and atmospheric stress-crop yield relationships for economic analysis.* Technical Bulletin 140. Oklahoma Agricultural Experiment Station, Stillwater.

27. Matanga, G. B., and M. A. Marino. 1979. *Irrigation planning. 1. Cropping pattern.* Water Resources Research 15(3): 672-678.

28. Meyer, G. E. 1979. *Simulation of moisture stress effects on soybean yield components in Nebraska.* Paper 79-4522. American Society of Agricultural Engineers, St. Joseph, Michigan.

29. Moore, C. V. 1961. *A general analytical framework for estimating the production function for crops using irrigation water.* Journal of Farm Economics 63: 876-888.

30. Morgan, T. H., A. W. Biere, and E. T. Kanemasu. 1980. *A dynamic model of corn yield response to water.* Water Resources Research 16(1): 59-64.

31. Musick, J. T., and D. W. Grimes. *Water management and consumptive use by irrigated grain sorghum in western Kansas.* Technical Bulletin 113. Kansas Agricultural Experiment Station, Manhattan.

32. Nebraska Natural Resources Commission. 1981. *Summary of the Nebraska research for the six-state High Plains Ogallala Aquifer study.* Lincoln.

33. Palacios, E. V. 1981. *Response functions of crop yield to soil moisture stress.* Water Resources Bulletin 17(4): 699-703.

34. Rochester, E. W., and C. D. Busch. 1972. *An irrigation scheduling model which incorporates rainfall predictions.* Water Resources Bulletin 8(3): 608-613.

35. Roeder, L. F. 1981. *Limited irrigation crop selection: A linear programming model.* M.S. thesis. Kansas State University, Manhattan.

36. Rydzewski, J. R., and S. Nairizi. 1979. *Irrigation planning based on water deficits.* Water Resources Bulletin 15(2): 316-325.

37. Shaw, R. H. 1980. *Modeling crop yields using climatic data.* In Proceedings, International Workshop on the Agroclimatological Research Needs of the Semi-Arid Tropics. International Crop Research Institute for the Semi-Arid Tropics, Hydenabad, India.

38. Stewart, J. I., and R. M. Hagan. 1969. *Predicting effects of water shortage on crop yield.* Journal of the Irrigation and Drainage Division, American Society of Civil Engineers 95: 91-104.

39. Stewart, J. I., R. D. Misra, W. O. Pruitt, and R. M. Hagan. 1975. *Irrigating corn and grain sorghum with a deficient water supply.* Transactions, American Society of Agricultural Engineers 18(2): 270-280.

40. Stewart, J. I., and R. M. Hagan. 1973. *Functions to predict effects of crop water deficits.* Journal of the Irrigation and Drainage Division, American Society of Civil Engineers 99: 421-439.

41. Tscheschke, P., J. R. Gilley, T. Thompson, and P. Fishbach. 1978. *IRRIGATE - A scheduling model.* Agricultural Engineering 59(1): 45-46.
42. U.S. Department of Agriculture. Various issues. *Agricultural prices.* Washington, D.C.
43. U.S. Department of Agriculture. Various issues. *Agricultural statistics.* Washington, D.C.
44. Watts, D. G., and R. J. Hanks. 1978. *A soil-water-nitrogen model for irrigated corn on sandy soils.* Soil Science Society of America Journal 16(2): 257-262.
45. Yaron, D., E. Bresler, H. Bielorai, and B. Harpinist. 1980. *A model for optimal irrigation scheduling with saline water.* Water Resources Research 16(2): 257-262.
46. Zavaleta, L. R., R. D. Lacewell, and C. R. Taylor. 1980. *Open-loop stochastic control of grain sorghum irrigation levels and timing.* American Journal of Agricultural Economics 62(4): 785-792.
47. Zovene, J., O. Buller, and J. Steichern. 1982. *Model to evaluate conservation, design, and economic feasibility of supplemental irrigation systems in the sub-humid region.* Kansas Water Resource Research Institute, Manhattan.

7

Agricultural Impacts on Environmental Quality

John A. Miranowski

Agricultural production involves the manipulation of the natural environment to increase the output of food and fiber. There are two primary environmental consequences of such activities. First, these productive activities create externalities, such as sediment, nutrient, and pesticide residuals entering upland, wetland, and aquatic environments. Second, cropland is also capable of producing other products, such as wildlife, but agricultural activities frequently destroy or at least reduce the productivity of wildlife habitat by altering upland areas and draining wetlands. Because wildlife is a common property resource, the farmer cannot capture the benefits of improved wildlife productivity, especially under agricultural conditions typical of the North Central Region.

The current and future impacts of the region's agriculture on environmental quality are accentuated by structural changes occurring in the sector. Farms, fields, and machinery are growing larger, accompanied by a decline, and in many cases elimination, of fence rows, windbreaks, wetlands, and grassed waterways. Second, relative energy prices have risen, causing a decline in the number of tillage operations performed and an increase in pesticide use, especially herbicides. Third, strong export demand and higher grain prices in the 1970s resulted in expansion of the region's cropland base onto more fragile and erosive lands and the use of larger quantities of commercial fertilizer. Finally, both to expand the cropland base and to reduce the risk associated with variations in the natural weather patterns, irrigation has expanded in the region.

It should be noted that the emerging impacts of agricultural trends can be both positive and negative. For example, increasing field size and declining wildlife cover will reduce wildlife habitat quality. Reduced tillage operations, however, may leave more food and cover to benefit wildlife. Likewise, bringing more cropland under cultivation may increase soil erosion and sedimentation problems, but reduced tillage on a larger share of cropland may reduce total erosion. Thus, on balance, it may only be possible to indicate the directional impacts of specific trends, but say little about the overall magnitude of such changes.

Measuring Agriculture's Impacts on the Environment

Two major problems arise in measuring the emerging environmental impacts of agriculture. First, no one has a "crystal ball." Projections of demand and supply for agricultural commodities, as well as the structural adjustments in the agricultural sector, are filled with uncertainty. Yet sectoral trends and adjustments are tied closely to these factors. Second, the physical and biological information base necessary to link the agricultural sector to environmental deterioration or improvement is lacking.[1] Even though agricultural residuals and impacts of farming practices may, in a few cases, be traced to the surrounding environment, concrete measures of the physical and biological impacts are frequently unavailable. Also, even if the environmental impacts can be measured, economists have not been overly successful in placing monetary values on these impacts. Such monetary information is essential in trading off the added benefits and costs to determine the socially optimal level of environmental quality or, alternatively, determining the economic impact of emerging changes in agricultural activities.

Given the current state of the art in measuring the impacts of agriculture on the environment, the following discussion will indicate possible sources of physical and biological accounting information, possible techniques for economic valuation, and several empirical studies that have attempted to measure the environmental impacts of agriculture in the North Central Region.

Physical and biological surveys lack the interdisciplinary input needed for an economic evaluation of environmental costs. Moreover, given the disparate nature of the data from these surveys, it is generally impossible to establish relationships among the residuals created by agricultural activities, the deposition or transport of these residuals into alternative environmental media, and the environmental impacts of these concentrations on health, wildlife, recreation, and aesthetics. There is a general lack of time series data or even periodic cross sections. Data generated by the 1977 National Resource Inventory (NRI) cannot be compared with

---

[1]See chapter 3, "Water Quality: A Multidisciplinary Perspective," page 36.

earlier soil resource surveys, such as the 1967 Conservation Needs Inventory. Fortunately for future research endeavors, the NRI was repeated in 1982 and supposedly will be performed every five years thereafter. The 1980 Resource Conservation Act evaluation (20), which provided an extremely useful summary of the available information on soil, water, and related resources, is to be repeated periodically. Current funding cuts, however, jeopardize the breadth and possibly the existence of the next evaluation.

Wildlife surveys are sometimes conducted to establish counts or supply, but little has been done to establish wildlife demand. As Hoekstra and associates (11) indicated, "Available information is very meager on supply and demand for wildlife and fish." Likewise, data collection activities of the U.S. Environmental Protection Agency and the U.S. Public Health Service have not been amenable to analysis of the impacts of agricultural activities on health and environmental quality. More research and better data collection are required. Given existing data and knowledge, present evaluations are limited to micro studies of isolated situations with suitable data. Such opportunities are not plentiful. Nor are the results necessarily capable of being generalized.

The available alternatives for measuring the economic value of environmental impacts are somewhat limited. The more acceptable alternatives include surveys, travel-cost studies, and property-value studies. Although somewhat tedious, the cheapest way to collect data on individuals' valuation of water quality benefits is to query a sample of users or potential users regarding their willingness to pay for a given water quality improvement (8). Economists tend to be somewhat skeptical of surveys because of the possibility of strategic behavior and because of the hypothetical nature of survey questions. Despite the heat generated in the economic literature, there is little evidence that individuals do employ strategic behavior and that significant bias is introduced by well-formulated hypothetical questions. More sophisticated survey techniques—contingent valuation, for example—are being developed to overcome the shortcomings of earlier surveys.

Travel-cost studies were used initially to measure the demand for recreational facilities (3). Visits to a particular recreation facility were related to price, i.e., the travel cost to the facility or distance, and other socioeconomic variables. This approach has been extended to account for differences in water quality, which are entered as right-hand-side variables. As Feenberg and Mills (8) stated, "If visits increase as water quality improves at the site, then it appears reasonable to assume that the gain in consumer surplus resulting from the improved water quality is related to the increased area under the demand curve for visits."

Intuitively, the property value or hedonic price studies compare two properties that are identical in all respects except for surrounding water quality. Presumably, property users should be willing to pay more for property associated with cleaner water, all other things equal. Moreover, the difference in value to the purchaser should be equal to the discounted value of the net benefits of cleaner water. There are a number of problems with using property-value studies to value the impacts of differences in environmental quality (*8*). These problems are accentuated in the case of water quality differences attributable to agricultural activities, given the dispersed nature of the impacts and the difficulties in attribution. It is probably safe to conclude that all of these analytical techniques, some more than others, will encounter serious obstacles in valuing the impacts of nonpoint-source pollution.

Isolated empirical studies have attempted to measure, and in a few cases value, the environmental impacts of agriculture in the North Central Region. A multiple-objective study (*2*) of the trade-offs between sediment damage and farm production costs for the Iowa River watershed reported a marginal sediment damage cost of $5.30 per ton for sediment delivered to the Coralville Reservoir. A study (*13*) of the Hambaugh-Martin watershed in Illinois placed sediment damage costs somewhere between $3.18 and $2.38 per ton as the discount rate varied from 2 to 8 percent.

The impact of erosion control policies on wildlife habitat located on private land has been investigated using a habitat indexing procedure (*16*). This approach is not suitable for valuing the habitat adjustments, but rather measures changes in wildlife habitat quality. A more sophisticated approach, the Habitat Evaluation Procedure, is under development by the U.S. Fish and Wildlife Service. The procedure is designed to value changes in wildlife habitat. Earlier wildlife valuation studies are unsuitable for assessing the impact of agriculture on wildlife habitat and environmental quality. An exception may be a recent study (*14*) that attempts to isolate the determinants of duck hunter participation along the Mississippi flyway. It was estimated that a 10 percent reduction in waterfowl habitat in the flyway has a discounted present value of $108 million. Although the researchers' purpose was not to assess agriculture's encroachment on flyway habitat, the technique may be suitable for assessing such impacts when appropriate data are available.

Therefore, given the current limitations on impact measurement and valuation, discussion of the environmental consequences of agriculture in the North Central Region will be primarily descriptive.

Agricultural Trends and Environmental Consequences

Analysis of agriculture in the North Central Region is complicated by the dynamic nature of output and input levels and the economic structure in the sector. Furthermore, as relative resource prices change and new technologies are developed, the sectoral adjustments are neither unidirectional nor easily predictable with current knowledge. For example, the impacts of relative energy price increases on the derived demands for labor, land, chemical, and machinery inputs are subject to different interpretations (5, 15).

Another complication is that the emerging impacts of agriculture on environmental quality need to be divided into short-run and long-run impacts. Current or short-run impacts are already occurring or will soon be realized and thus are easier to assess. Long-run impacts are far more uncertain and speculative. Assessments of long-run impacts must be viewed with caution and healthy skepticism.

Despite these problems, recent studies, including the National Agricultural Land Study (17), Crosson and Brubaker (4), and the National Research Council (18), have made projections of agricultural trends and environmental impacts for the 48 contiguous states. My discussion will draw heavily upon the studies by Crosson and Brubaker and the National Research Council. Crosson and Brubaker go into more detail on agricultural projections, while the National Research Council study places more emphasis on the wildlife habitat impacts. My assessment focuses on the 12-state North Central Region as opposed to the 48 contiguous states, which were considered in the other studies. Primary emphasis will be given to soil loss, fertilizer use, pesticide use, drainage of wetlands, and irrigation.

Soil Loss

Soil Loss Trends. Presently, the soil loss issue is receiving widespread public attention. Higher food and feed grain prices in the 1970s encouraged an expansion of the cropland base in the North Central Region. The new acreages brought under cultivation were frequently more fragile and prone to more serious erosion when used in row crop production. Table 1 indicates the magnitude of recent cropland expansion in the North Central Region. Even if the new lands brought under cultivation were eroding at the same average rate as existing cropland, the magnitude of the cropland increase would make a major contribution to total soil loss and to sediment delivery to streams and lakes.

A more detailed state breakdown for the sheet and rill erosion problem is presented in table 2. Based on these data, Iowa and Missouri had more than 45 percent of their cropland eroding at a rate exceeding five tons per acre in 1977, while Illinois had 37 percent and Indiana had 28 percent. Although there are no estimates of sediment delivery to water resources in these states, the receiving or surface water areas (Table 3) in the states are quite limited, implying a heavy sediment load per unit area relative to that of the other states in the North Central Region.

Given the current downward trend in grain prices and allowing a reasonable adjustment lag, a short-run reduction in cropland and row crop acreage is expected as the least profitable (and more erosive) croplands are removed from crop production. These lands will revert to pasture or will be abandoned. Thus, both total and average erosion should decline in the next few years as less productive lands are withdrawn from production. The long-range implications will depend upon the future growth in domestic and foreign demand for grain and the rate of growth in pro-

Table 1. Harvested cropland acres in the North Central Region in 1964 and 1978 (22).

State	Harvested Cropland Acres	
	1964	1978
	—— million acres ——	
Illinois	20.0	22.8
Indiana	10.3	11.9
Iowa	20.0	23.8
Kansas	18.2	19.1
Michigan	6.7	7.0
Minnesota	17.5	19.2
Missouri	11.1	12.6
Nebraska	15.2	16.4
North Dakota	17.7	19.1
Ohio	9.3	10.3
South Dakota	14.4	13.9
Wisconsin	9.0	10.0
North Central Region	169.4	186.1

Table 2. Sheet and rill erosion on cropland in the North Central Region, 1977 (20).

State	Sheet and Rill Erosion (tons/acre/year)					
	Less than 5 Tons per Acre per Year		5 to 13.9 Tons per Acre per Year		14 or More Tons per Acre per Year	
	1,000 Acres	Percent of State	1,000 Acres	Percent of State	1,000 Acres	Percent of State
Illinois	15,111	63.4*	6,422	26.9	2,303	9.7
Indiana	9,566	71.8	2,846	21.4	908	6.8
Iowa	14,452	54.7	7,018	26.6	4,961	18.8
Kansas	23,615	82.0	4,272	14.8	919	3.2
Michigan	8,481	89.4	695	7.3	308	3.2
Minnesota	20,594	89.9	1,827	8.0	495	2.2
Missouri	7,668	52.6	3,912	26.8	2,993	20.5
Nebraska	15,966	77.1	2,528	12.2	2,205	10.7
North Dakota	24,817	92.2	1,924	7.1	172	0.6
Ohio	9,916	84.3	1,499	12.7	347	3.0
South Dakota	16,222	89.3	1,580	8.7	354	1.9
Wisconsin	9,429	80.3	1,804	15.3	508	4.3
North Central Region	175,837	76.9	36,327	15.9	16,473	7.2

*Percentage of total cropland for each state.

Table 3. *Surface water distribution in the North Central Region* (21).

State	Inland Water Area (square miles)
Ohio	204
Indiana	102
Illinois	525
Michigan	1,399
Wisconsin	1,688
Minnesota	4,779
Iowa	247
Missouri	640
North Dakota	1,385
South Dakota	1,091
Nebraska	705
Kansas	208
Total	12,973

Table 4. *Conservation tillage in the North Central Region.* *

	Percentage in Conservation Tillage	Percentage Amenable to Conservation Tillage
Lake States	18	
Northern Plains	34	
Corn Belt	33	
Ohio	8	48
Indiana	23	53
Illinois	28	66
Iowa	39	76

*The regional estimates are for 1981 and were developed by Crosson and Brubaker (4) based on *The No-Till Farmer* estimates. The state estimates are for 1979 from the National Research Council (18) and were reported by the Soil Conservation Service and Cosper.

ductivity. If productivity growth of these crops exceeds the rate of demand growth or if other inputs become relatively cheaper, then the cropland base can be expected to continue declining with the soil erosion problem becoming less serious.

A positive factor in current agricultural production is the spread of conservation tillage. Reduced tillage operations leave more residue on the soil surface, reducing the ability of water and wind to remove soil particles. Table 4 provides information on current conservation tillage use in the North Central Region and the potential for future adoption of the technology in four states. At a minimum, the doubling of conservation tillage is technically, if not economically, feasible in all four Corn Belt states. All indications are that conservation tillage will continue to expand as producers become better informed and as improved technologies become available. However, technical obstacles remain, especially under less than good management conditions.

The spread of conservation tillage has been spurred by rising relative energy prices in the past decade. In an effort to reduce energy costs, farmers are substituting more sophisticated machinery and chemical weed control for additional trips across the field. The relative energy price trend will likely continue upward over the longer term, further spurring the adoption of conservation tillage and, to a lesser extent, no-till techniques. Rising energy prices will encourage the adoption of less erosive practices on the land under cultivation (15).

In summary, the combination of depressed grain prices and relatively higher energy prices should contribute substantially to a reduction in soil loss within the next decade. Yet a caveat is in order. Land is a substitute for energy.[2] If relative energy prices continue to rise, *ceterus paribus,* then cropland may substitute for energy. Productivity per acre would drop, but more cropland acres would come into production, offsetting some of the potential reductions in soil loss.

Soil Loss Impacts. The off-site impacts or externalities from soil loss include reduced reservoir capacity, reduced recreational opportunities, fish and wildlife habitat deterioration, and increased siltation of waterways, harbors, and ditches. Suspended solids are a nonpoint-source pollutant affecting more than half of the Great Lakes hydrologic basins and 80 percent of the North Central basins (Table 5). In many cases, these suspended solids are the result of soil losses.

What are the emerging impacts of current trends in agricultural production? The short-run impacts point in the direction of reduced soil loss and improvements in water quality and aquatic habitat. Upland habitat will also improve as the spread of reduced tillage leads to improved cover and food supplies for wildlife and as habitat diversity is improved by a reduction in row crop acreage and an increase in pasture or grassland.

Projections of long-range impacts are more pessimistic.[3] The studies cited earlier (4, 17, 18) project a long-term growth in feed and food grain demands and continued expansion of the cropland base. Table 6 presents estimates of the potential cropland that is currently in other uses. With current technology, an increase in the cropland base of less than 20 per-

---

[2]Miranowski, John A., and Edward K. Mensah. 1979. "Derived Demand for Energy in Agriculture: Effects of Price, Substitution, and Technology." Paper presented at the Agricultural Economics Association Annual Meeting, July 29-August 1, Washington State University, Pullman.

[3]Unanimous agreement on the longer term agricultural sector adjustment does not exist. See Chapter 5, "The Future of Agriculture in the North Central Region," for a more optimistic picture. If the long-range projections by Swanson and Heady are employed, the long-range environmental impacts would be less serious.

cent appears feasible. Given the characteristics of this potential cropland, a 20 percent increase would have significant environmental costs.
Using similar estimates of cropland potential, Crosson and Brubaker (4)
project a doubling of soil erosion by 2010. Even the spread of conservation tillage technology cannot control or sufficiently moderate the in-

Table 5. Percentage of hydrologic basins affected by
nonpoint-source pollution in the North Central and
Great Lakes Regions (18).

	Basins Affected by Pollution	
	Great Lakes*	North Central
Number of basins	41	35
	——— % ———	
Source of pollution		
Urban runoff	54	54
Construction	7	6
Hydrologic modification	2	3
Silviculture	15	6
Mining	41	40
Agriculture	59	89
Solid waste disposal	15	9
Individual disposal	39	29
Bacteria	51	69
Oxygen depletion	54	66
Type of pollution		
Nutrients	44	63
Suspended solids	56	80
Dissolved solids	27	51
pH	37	20
Oil and grease	20	0
Toxics	34	51
Pesticides	15	37

*EPA designation of Great Lakes and North Central
Regions.

Table 6. Use of agricultural lands, 1977, and potential
conversion to cropland by source in North Central
Region (17, 20).

	Current Cropland	Potential Cropland
	— million acres —	
Cropland	228.7	228.7
Pasture	41.5	18.8
Rangeland	71.1	12.6
Forest	69.2	6.9
Other	7.2	2.0
Total potential cropland		268.7

creased soil loss problems. Likewise, the habitat destruction impacts would be sizeable.

Fertilizer Use

Fertilizer Use Trends. Expansion of the cropland base in the North Central Region, combined with changing relative prices, has led to a 4.5-fold increase in plant nutrient use over the last two decades. Although not illustrated in table 7, a far greater increase occurred in nitrogen use. In addition, following the shift of comparative advantage in feed grain production to the North Central Region (*19*), the share of plant nutrients used in the region has increased from 38 percent in 1960 to 58 percent in 1979. Not only did the total quantity expand, but the quantity per acre of cropland increased. Although quantitative data are unavailable, the amount of nutrients delivered to surface and groundwater likely increased as well.

Recent indications are that the combination of rising nutrient prices and declining commodity prices are leading to more judicious use of fertilizer nutrients in the short run. The trend even indicates modest declines in per-acre use rates. Although the current world anhydrous ammonia glut and a depressed national economy cloud the issue somewhat, total deregulation of natural gas prices as well as anticipated increases in other nutrient prices can be expected to encourage more efficacious use of fertilizer nutrients in the long run as well. For example, better placement and improved timing of nutrient application may allow the producer to maintain productivity while reducing actual application levels. Once again, relative prices coupled with technological factors will be the long-term driving forces. If nutrient residuals in the environment are perceived as a more serious problem, government intervention could also be a key factor.

Fertilizer Use Impacts. Fertilizer nutrients can have a variety of impacts on environmental quality. Nitrogen in water can be toxic to humans and animals. Nitrogen and phosphorus both can stimulate plant

Table 7. Shifting regional use of plant nutrients, 1960, 1970, and 1979 (10).

	Nutrient Use							
	East-North Central		West-North Central		North Central		United States	
Year	1,000 Tons	Percentage	1,000 Tons	Percentage	1,000 Tons	Percentage	1,000 Tons	Percentage
1960	1,707	23	1,175	16	2,882	39	7,464	100
1970	3,962	25	4,448	28	8,411	53	16,068	100
1979	6,114	27	6,873	31	12,987	58	22,569	100

growth in water bodies, leading to accelerated eutrophication, which causes declines in dissolved oxygen levels. Fish and recreational use of water resources both may suffer as a result. Nitrogen is transported to the stream in runoff water and water leaching through drainage tile. Most phosphorus is transported to water bodies in a form bound to soil particles. Thus, controlling soil erosion will control most pollution from agricultural phosphorus.

Adoption of conservation tillage practices will alter the nutrient residual problem (4). Leaving more crop residue on the soil surface reduces water runoff and thus soil erosion, but it increases water percolation and tile drainage flows. Because phosphorus binds to soil particles, phosphorus residuals entering streams and lakes should decline, all other things equal. At the same time, nitrogen residuals, which are more concentrated in drainage flows, will likely increase in streams and lakes.

In the North Central Region, human and animal health problems associated with nitrogen residuals are limited to occasional periods following heavy spring rainfall when the Public Health Service standard of 10 parts per million of nitrate-nitrogen is exceeded. Generally, these concentrations are not widespread and are probably not reason for serious short-run concern.

Recently, irrigated areas in the Nebraska Sandhills have reported nitrate-nitrogen concentrations in groundwater of 20 parts per million (7), a level that does warrant concern for the short-term. The long-range problem is likely to become more serious if irrigation in the region continues to expand. Although good irrigation management is capable of controlling the problem, such action is not assured, especially if significant irrigation expansion continues.

The seriousness of eutrophication impacts are more complex. Phosphorus is often the limiting factor in plant growth in water bodies. Phosphorus residuals originate from many sources, but detergents are a major culprit. The available evidence suggests that eutrophication is more severe around heavily populated areas and that fertilizer residuals contribute less than other sources. From an efficiency perspective, if phosphorus residuals are a problem, abatement should not necessarily be undertaken in proportion to contribution or relative to magnitude of contribution, but rather on the basis of least-cost abatement. Unfortunately, information is neither readily available on the comparative costs of abatement nor on the relative contributions of different phosphorus sources in the North Central Region. Table 5 indicates that nutrients do affect a number of hydrologic basins, but it does not indicate which nonpoint source is responsible.

The long-run projections by Crosson and Brubaker (4) indicate an 87 percent increase in nitrogen and a 68 percent increase in phosphorus applications for the United States by 2010. Even though they project smaller increases for the North Central Region and do not predict serious nutrient pollution problems, except in the Nebraska Sandhills, there is a high probability that nutrient residual problems from agriculture will emerge by 2010, if not before, in the North Central Region. The long-range relative changes in regional crop production shares on which this outlook is based depend upon short-term shifts, not adjustments in long-term comparative advantage. An alternative scenario would assume more intensive use of agricultural inputs in the North Central Region to satisfy growing demand. Associated with such a scenario are rising environmental costs for the region.

Pesticide Use

Pesticide Use Trends. From an environmental perspective, pesticides present another area of concern. As table 8 indicates, insecticide use in the North Central Region increased slightly from 1971 to 1976, but herbicide use increased significantly. Preliminary data for 1980 indicate that these trends have continued (*6*).

The rapid growth in herbicide use is a matter of environmental concern. Expanding use of conservation tillage will further increase the use of herbicides. Although some of the growth of herbicide use in the North Central Region may be tempered by short-term reductions in the cropland area and by the growth in integrated crop and pest management, an upward trend is predicted in both acres treated and use rates per acre. Current weed control problems in reduced tillage fields also may necessitate further increases in application rates.

From an environmental perspective, reduced tillage may not always be the panacea that some believe it to be. Besides the expected increase in herbicide use per acre with this system, evidence is beginning to accumulate that other pest problems may increase over time with reduced tillage. Surface residue may harbor insect and disease organisms, permitting

Table 8. Quantities of pesticides by type of pest control for North Central States in 1971 and 1976 (1, 7).

Category and Year	Pesticide Use		
	Lake States	Corn Belt	Northern Plains
	——— 1,000 lbs. ———		
Insecticides			
1971	4,752	20,421	7,547
1976	5,201	15,738	11,013
Herbicides			
1971	36,707	81,610	28,231
1976	44,039	155,277	43,219
Fungicides			
1971	2,469	5,297	539
1976	2	16	-

them to carry over from one crop season to the next. New pest problems may migrate into the North Central Region from southern states if a more suitable winter habitat is provided. Conventional tillage technologies, such as moldboard plowing, have provided a form of sanitation and pest control that encouraged their initial adoption and continued use. Although there are advantages in soil loss control associated with conservation tillage, a shift from clean-till cultivation may have long-run environmental costs, especially with the increased use of pesticides.

Pesticide Use Impacts. On balance, pesticides do not appear to present the environmental threat that other agricultural residuals in the North Central Region do. Concentrations of pesticides in water have declined and the negative water quality impacts have been reduced in recent years. The new generation of pesticides, especially insecticides, break down rather quickly and do not constitute a serious hazard to the aquatic environment unless misapplied or applied immediately preceding a heavy rainfall. With the exception of toxaphene, the other organochlorine insecticides are banned or restricted to special situations.

The organophosphorus and carbamate insecticides, although sometimes far more toxic, have short half-life periods and do not persist long in the environment. Generally, herbicides are also short-lived and, with the exception of paraquat, have low human toxicity. In addition, the spread of conservation tillage, although encouraging higher application rates, reduces runoff and the residual load entering streams.

To avoid dismissing pesticides too lightly, it is important to note that solid scientific knowledge is lacking on the breakdown products of these chemicals. The jury is still out on their impacts on human health (mutagenic and carcinogenic effects, for example) and the environment. Less is known about the environmental hazards accompanying herbicide use. Therefore, because of the high levels of herbicide use in the North Central Region, along with possible increased use, careful study of the consequences is necessary.

Given the lack of evidence to the contrary, neither the short-term nor long-term impacts of pesticides used in the North Central Region appear to be alarming. Use of integrated crop and pest management will substitute for a portion of the potential increased use of pesticides. Moreover, the prospects of more environmentally compatible compounds should offset many of the potential negative long-range impacts. Other agricultural regions of the nation may not be so fortunate.

Wetland Drainage

Wetland Drainage Trends. Reliable regional time series data on wetland drainage are not available. However, not all of the expansion of cropland in the last decade in the North Central Region occurred on erosive, hilly land. Some of the new cropland was acquired by draining wetlands. Unlike the cultivation of steep lands, which may again revert to pastures, once wetlands are drained and pressed into cultivation the change is generally irreversible.

Intensive study of the prairie pothole area of Minnesota, North Dakota, and South Dakota indicates that substantial drainage activities are continuing (*18*). About 35,000 acres per year were drained in the area during the late 1960s (*12*). Similarly, a 40 percent reduction in wetland area occurred in southwestern Minnesota from 1964 to 1974 (*18*). Although export demand was not a significant factor during most of this period, subsidization of drainage by the Soil Conservation Service probably was important. It is believed that the rate of drainage increased during the 1970s, but actual documentation is not available.

Existing state laws have slowed but not stopped drainage activities. The easy areas to drain have been brought under cultivation, but improvements in drainage technology are now making it feasible to drain deeper wetland areas. It is probably safe to assume that drainage will continue, but at a more moderate rate than in the past.

Wetland Drainage Impacts. Wetlands play an extremely important biological role in maintaining environmental integrity. First, wetlands are crucial habitat for both resident and migrant wildlife species, providing water, food, and cover. Second, wetland areas serve as traps for sediment and nutrients from agricultural production. Third, the hydrologic function of wetlands is instrumental in maintaining base stream flows and other aquatic environments (*18*). Drainage, tiling, and channelization investments will continue to destroy these functional roles of wetlands. Some wetlands will be drained to dry out uplands and remaining wetlands will be enlarged, destroying their marsh habitat role (*18*).

At the same time, the increased agricultural activities in the drained areas will increase the fertilizer and pesticide residuals entering the remaining marshes, as well as the silt load, thus reducing water quality in these wetlands. If these agricultural activities are accompanied by irrigation, substantial increases in residual loading will result, and water tables may be drawn down in the remaining wetland areas. Given the declining supply of inland wetland areas and presumably a growing demand for the wildlife products of wetlands, further losses will impose significant costs on society.

Irrigation

Irrigation Trends. Irrigation has been an important agricultural input in Nebraska and Kansas. Significant increases in irrigation occurred in all North Central Region states from 1964 to 1978 (Table 9). Total irrigated acreage increased 2.8 times. Substantial increases in acreage under supplemental irrigation have occurred in the Corn Belt, primarily in Missouri, and in the Lake States. Rising energy prices, falling water tables, and lower wheat and corn prices are slowing the rate of expansion and in some areas reducing irrigated acreage. In the long-run, however, continued expansion of irrigation in the North Central Region is expected to further reduce stream flows, lower water tables, reduce wetland areas, and diminish water quality in the affected areas.

Irrigation Impacts. Although Frederick (9) concluded that "salinity is the most pervasive environmental problem stemming from irrigation in the United States," the wildlife and recreational impacts are more serious in the North Central Region. Presently, salt concentrations do not threaten the viability of most irrigated agriculture in the region, even in the Northern Plains States.

In the Northern Plains, initial increases in irrigated agriculture were beneficial to wildlife because of increased food supplies (18). Yet continued expansion of irrigation can be expected to have negative impacts. As dryland cropland and rangeland are converted to row crops, edge, cover, and vegetative diversity decline. Declining water tables and stream flows, coupled with the lack of winter cover, will have detrimental impacts on wildlife habitat.

In the Corn Belt and the Lake States, supplemental irrigation will require more intensive crop production management, including increased land leveling, clearing, drainage, and chemical use, and reduced cropping diversification. Although many of these adjustments have occurred

Table 9. Irrigated acreages in the North Central Region, 1964 and 1978 (22).

State	Irrigated Acres	
	1964	1978
	— 1,000 acres —	
Illinois	14	131
Indiana	17	76
Iowa	22	101
Kansas	1,004	2,607
Michigan	49	227
Minnesota	18	272
Missouri	59	344
Nebraska	2,169	5,541
North Dakota	50	144
Ohio	17	26
South Dakota	130	345
Wisconsin	62	236
North Central Region	3,611	10,050

in the absence of irrigation, the long-range impacts will bring a further reduction in wildlife habitat and in consumptive and nonconsumptive recreational opportunities in the affected areas. Increased irrigation will reduce base flows of streams and stream habitats. Reductions in groundwater recharge will reduce base flows and dry up wetlands. For example, groundwater irrigation in Wisconsin has destroyed the trout fishing on some streams in that state (18). Expanded irrigation in Michigan could have similar long-range impacts.

Conclusions and Outlook

In the short run the impacts of agriculture in the North Central Region may lead to improvement in environmental quality. Given current prices for food and feed grains, producers have an incentive to reduce their cropland acreages and to forego expansion of irrigation and drainage activities. These adjustments should be accompanied by reductions in total soil loss and in total chemical use. In addition, if the energy price trend again turns upward in the near future, chemical and water use per acre may decline and the spread of conservation tillage should be encouraged. All of these adjustments would have positive impacts on environmental quality.

Although more uncertain, the long-range outlook for the environmental impacts of agriculture in the North Central Region is more pessimistic. Most projections indicate a growing foreign demand for agricultural products and increased incentives for agriculture in the region to increase output. Unless significant technological improvements are forthcoming, increased output will require both more intensive use of inputs and a larger cropland base. The pressures on environmental quality will increase if more chemicals, water, and cropland are pressed into service. Some of these problems may be mitigated by the spread of conservation tillage, more efficient use of fertilizers, adoption of integrated crop and pest management, rising energy prices, and legislation protecting environmental quality.

The major environmental quality impacts of agriculture in the North Central Region, including those caused by soil loss, chemical use, wetland removal, and irrigation, have been enumerated. Unfortunately, the knowledge and data necessary to establish quantitative connections between agricultural activities and environmental quality are seldom available. Furthermore, even if these connections could be established, significant problems may be encountered in valuing the measured impacts to arrive at monetized environmental costs. Yet to arrive at enlightened decisions, information is needed on the dollar values of trade-offs.

It is unnecessary and inefficient to maintain the same standards of environmental quality in all areas of the nation or the North Central Region. In addition to the supply of environmental quality, the demand for environmental quality is an important factor in determining the socially desired level. Because demand and supply vary by area, so do the optimal levels. Some areas may be demanded for recreational purposes or for waterfowl flyways and others may not. Likewise, the opportunity costs of supplying water quality in some areas may be much higher than in other areas. Ultimately, these factors should enter into environmental quality decisions.

A movement is currently underway to shift decision-making responsibilities from the federal government to the states. In the case of water quality and quantity, such a move may increase economic efficiency in many cases, but only if adequate information on costs and benefits is available to the state authorities. Given the current paucity of such data, it is doubtful if much economic gain can be realized. Only if the necessary research information can be assembled will substantial gains be made.

REFERENCES

1. Andrilenas, Paul A. 1974. *Farmers' use of pesticides in 1971—quantities.* Agricultural Economic Report No. 252. Economic Research Service, U.S. Department of Agriculture, Washington, D.C.
2. Boggess, William, John Miranowski, Klaus Alt, and Earl Heady. 1980. *Sediment damage and farm production costs: A multiple-objective analysis.* North Central Journal of Agricultural Economics 2(2): 107-112.
3. Clawson, Marion, and Jack L. Knetsch. 1966. *Economics of outdoor recreation.* John Hopkins Press, Baltimore, Maryland.
4. Crosson, Pierre R., and Sterling Brubaker. 1982. *Resource and environmental effects of U.S. agriculture.* Resources for the Future, Inc., Washington, D.C.
5. Duncan, Marvin, and Kerry Webb. 1980. *Energy and American agriculture.* Federal Reserve Bank of Kansas City. Kansas City, Missouri.
6. Eichers, Theodore R. 1981. *Farm pesticide economic evaluation, 1981.* Agricultural Economic Report No. 464. Economics, Statistics, and Cooperatives Service, U.S. Department of Agriculture, Washington, D.C.
7. Eichers, Theodore R., Paul A. Andrilenas, and Thelma W. Anderson. 1978. *Farmers' use of pesticides in 1976.* Agricultural Economic Report No. 418. Economics, Statistics, and Cooperatives Service, U.S. Department of Agriculture, Washington, D.C.
8. Feenberg, Daniel, and Edwin S. Mills. 1980. *Measuring the benefits of water pollution abatement.* Academic Press, New York, New York.
9. Frederick, Kenneth D. 1982. *Water for western agriculture.* Resources for the Future, Inc., Washington, D.C.
10. Hargett, Norman L., and Janice T. Berry. 1981. *1980 fertilizer summary data.* Bulletin Y-165. National Fertilizer Development Center, Tennessee Valley Authority, Muscle Shoals, Alabama.
11. Hoekstra, Thomas W., Dennis L. Schweitzer, Charles T. Cushwa, Stanley H. Anderson, and Robert B. Barnes. 1979. *Preliminary evaluation of a national wildlife and fish data base.* In transactions of the Forty-Fourth North American Wildlife Conference. Wildlife Management Institute, Washington, D.C. pp. 380-391.
12. Horowitz, E. L. 1978. *Our nation's wetlands.* Council on Environmental Quality, Washington, D.C.
13. Lee, M. T., A. S. Narayanan, Karl Guntermann, and E. R. Swanson. 1974. *Economic*

analysis of erosion and sedimentation: Hambaugh-Martin watershed. AER-127. Department of Agricultural Economics, University of Illinois, Urbana.

14. Miller, John R., and Michael J. Hay. 1981. *Determinants of hunter participation: Duck hunting in the Mississippi flyway.* American Journal of Agricultural Economics 63(4): 677-684.

15. Miranowski, John A., Michael S. Monson, James S. Shortle, and Lee D. Zinser. 1982. *Effect of agricultural land use practices on stream water quality: Economic analysis, 15 October 1980-15 January 1982.* U.S. Environmental Protection Agency, Washington, D.C.

16. Miranowski, John A., and Ruth Larson Bender. 1982. *Impact of erosion control policies on wildlife habitat on private lands.* Journal of Soil and Water Conservation 37(5): 288-291.

17. National Agricultural Lands Study. 1981. *Final report.* U.S. Department of Agriculture, Washington, D.C.

18. National Research Council. 1982. *Impacts of emerging agricultural trends on fish and wildlife habitat.* National Academy Press. Washington, D.C.

19. Schultz, Theodore W. 1982. *The dynamics of soil erosion in the United States: A critical view.* Agricultural Economics Paper No. 82:8. University of Chicago, Chicago, Illinois.

20. U.S. Department of Agriculture. 1980. *Soil and Water Resources Conservation Act, 1980 appraisal.* Washington, D.C.

21. U.S. Department of Commerce. 1970. *Area measurement reports.* Bureau of the Census, Washington, D.C.

22. U.S. Department of Commerce. 1967 and 1981. *U.S. census of agriculture, 1964 and 1978.* Bureau of the Census, Washington, D.C.

8

Competition for Water, a Capricious Resource

K. William Easter, Jay A. Leitch, and Donald F. Scott

As Sir Alan Herbert's poem "Water" suggests, water takes different forms and usually is not located where man would wish it to be. Water has been transported great distances to make arid areas more livable. In the North Central Region of the United States, water quality, timing, and certainty of supply are much more important than quantity when considering the competitive uses of water. Heavy spring rains, for example, may only delay planting and be of little use during a dry summer. Water that is highly polluted is not available without treatment for domestic and industrial uses. Water in the form of precipitation will be available to the North Central Region only with a degree of uncertainty.

Because large interbasin water transfers appear to be unsound investments under any reasonable set of assumptions (*11, 17*), our focus here is on competitive uses of water available within the North Central Region. If fresh water becomes scarce in other regions, we can expect this to provide the North Central Region with a competitive advantage, not a basis to transfer large quantities of water.

The North Central Region is far from uniform in the competitive uses of water as one moves from west to east and north to south in the region. Irrigation, by far the largest use of water in the western area, is much less significant in the east. Competitive uses thus differ throughout the region. In addition, the source of water varies widely. Groundwater is dominant in states such as Kansas and Nebraska, while surface waters are of prime importance in the Lake States. Consideration of the competitive uses of water in the region must be in a context recognizing that the region has the largest area of fresh water of any region in the world.

Competition can assume a variety of forms. There may be competition among users for the same use, irrigation water for example. Or, there may be competition among uses, such as between irrigation and energy production or between recreation and navigation. Competition among geographic regions can occur when a watercourse extends through several political jurisdictions. Yet another form of competition is intertemporal, or through time: use of a groundwater stock today competes with future use. A final, general form of competition for water is indirect, when one use impairs another. This may occur when sulfides and chlorides pollute the air and return to earth as acid rain, ending up in lakes and streams. This use of watercourses as waste disposers competes with water-based recreation, which requires water that is clean and can support fish production.

To make our job manageable, we will focus on the following primary competitive water uses: (1) irrigation, (2) municipal and industrial, (3) recreation, (4) navigation, and (5) energy production.

Economic Framework

If we were dealing with a conventional input sold in competitive markets to produce alternative goods and services, we would expect the value of the input to be the same in each use at the margin. Cotton used to produce a shirt will have the same value per unit as that to produce curtains. With water, the marginal value in different uses can vary widely. This is the case for a number of reasons. First, there is no national market for water as there is for cotton. In cases where a market for water exists, it is often imperfect, includes subsidies, and is usually for only one use and in limited areas, for example, a market for water shares in an irrigation district. Second, the value of water varies by location, quality, and time of availability. Thus, water tends to be a less standardized commodity than most inputs. Third, water laws and institutions make it difficult, and in some cases impossible, to transfer water among uses. Consequently, industry can be paying in excess of $100 an acre-foot while irrigators are paying less than $10 an acre-foot.

Finally, nonconsumptive water uses and the potential future value of groundwater stocks make it difficult to place a value on water. Many uses of water do not preclude other water uses. For example, recreational use of water does not necessarily reduce its value downstream for uses such as transportation and recreation. In addition, part of the water used for irrigation will be used again downstream. Thus, for some uses we should value water like any other "private good" that is consumed; for other uses we should add the values in the different uses like a "public good" (whereby consumption by one person does not reduce its value to anyone else). The same water can be used for recreation, navigation, and power production.

The intertemporal question will be a dominant concern in the areas of extensive groundwater use, such as Kansas and Nebraska. The question is when and at what rate should groundwater be used. Currently, we rely on the private market to determine pumping or withdrawal rates. This ignores the common property nature of groundwater and the increased pumping cost imposed on neighboring farmers and on future generations.[1] In some states the externality created for other farmers, well interference, has been dealt with through legal restrictions. However, the "user costs" that the pumping of fossil water imposes on future generations has been ignored.[2]

Water for Irrigation

The competitive use of water for irrigation is concentrated in the western part of the North Central Region. Irrigation from surface water is important in the Dakotas, while groundwater irrigation has expanded rapidly in Kansas and Nebraska. Irrigation in these four Northern Plains states makes this region one of the most important irrigated areas of all farm production regions in the United States. In terms of irrigated areas, it has grown more rapidly than all other regions in the United States. The Northern Plains, with 10.6 million irrigated acres in 1977, ranks only behind the Pacific and Mountain Regions in irrigated cropland and ahead of the Southern Plains (Table 1). Center pivot technology has been one of the keys to this development, along with the high farm prices of the mid-1970s.

The Corn Belt and the Lake States Regions have limited irrigated acreage, about 1 million acres each (Figure 1). Most of this irrigated acreage is in the lower rainfall parts of the western Corn Belt and Lake States. Both groundwater and surface water sources are tapped to provide water. This supplementary irrigation has expanded rapidly, particularly after a couple of dry years, in areas with sandy or shallow soils.

Even in years with normal rainfalls, sprinkler irrigation in states such as Minnesota has proved profitable on sandy soils during periods of

[1]Groundwater that is overlain by surface lands owned by several individuals who also own the subsurface resources has many of the characteristics of a common property or fugitive resource. Any of the surface owners can pump the water no matter what impact it has on the neighbors. Surface owners will benefit from the groundwater only if they pump it before their neighbors do. Thus, there is no benefit to the individual surface owner who leaves the water in the ground.

[2]User costs are the present value of all future sacrifices associated with the use of a unit of groundwater. This includes foregone future uses, higher future pumping costs, and any adverse environmental costs (9).

good farm prices. However, for the deeper and heavier soils, irrigation seems to be more an insurance premium against dry years. Most farmers find more profitable ways, other than irrigation, to increase yields on heavier soils.

Currently, the competition for irrigation water is in the Northern Plains States and tends to be location specific. Irrigation uses the largest amount of water (from 1 to 2 acre-feet of water are applied annually per acre in much of the North Central Region; therefore, each 1,000 acres irrigated requires from 1,000 to 2,000 acre-feet of water per year). Where supply is a constraint it would not take a very large reduction in irrigation to satisfy other competitive uses. However, a few areas exist where the demands for other uses, such as energy production or industry, may require a reduction in irrigated acreage. This competition occurs only during certain times of the year or during drought years. Irrigated farms

Table 1. Irrigated and dryland agricultural land use by farm production region, 1977 (millions of acres) (4, 20).

Farm Production Region*	Irrigated Cropland	Dryland Cropland	Total Cropland
Northern Plains	10.6	83.9	94.6
Southern Plains	8.6	33.6	42.2
Mountain	15.2	27.1	42.2
Pacific	11.9	11.2	23.2
Total: 17 western states	46.4	155.8	202.2
Northeast	0.4	16.5	16.9
Appalachian	0.4	20.4	20.8
Corn Belt	1.1	88.8	89.9
Lake States	1.0	43.2	44.1
Southeast	2.4	15.1	17.5
Delta States	4.0	17.2	21.2
Total: 31 eastern states	9.3	201.2	210.4
National Total	55.7	357.2	412.9

*U.S. Department of Agriculture's farm production regions. The four western regions are defined as follows: Northern Plains includes Kansas, Nebraska, North Dakota, and South Dakota; Southern Plains includes Texas and Oklahoma; Mountain includes Arizona, Colorado, Idaho, Montana, Nevada, New Mexico, Utah, and Wyoming; Pacific includes California, Oregon, and Washington. The six Eastern regions are: Northeast includes Maine, New Hampshire, Vermont, Massachusetts, Rhode Island, Connecticut, New York, New Jersey, Pennsylvania, Delaware, and Maryland; Appalachian includes Virginia, West Virginia, North Carolina, Kentucky, and Tennessee; Corn Belt includes Ohio, Indiana, Illinois, Iowa, and Missouri; Lake States includes Michigan, Wisconsin, and Minnesota; Southeast includes South Carolina, Georgia, Alabama, and Florida; Delta States includes Mississippi, Louisiana, and Arkansas.

Figure 1. North Central Region.

in areas of extensive coal development are candidates for water use competition.

Other parts of the region will also face competition for irrigation water during drought periods. Again this will be location specific and primarily between municipal uses and irrigation. As residential areas continue to expand and households drill their own wells, they will compete with one another and with pumping for irrigation. The amount of acreage and water involved will still be small when compared with the total.

As irrigation expands in the Nebraska Sandhills, more competition will occur between agriculture and wildlife. Increased pumping may dry out wetlands and reduce streamflow. This means that certain wildlife species will lose habitat, and local water uses may be precluded. Reduced streamflow will also mean competition with downstream irrigators.

The key question for irrigation in the North Central Region involves diminishing groundwater aquifers. How rapidly should this stock groundwater resource be used? Currently, the only real regulation of pumping is pumping cost and the value of agricultural commodities produced. The current cost of energy relative to agricultural prices may slow expansion of well irrigation. Given potential future needs for groundwater, this may be a fortunate respite. A pause in groundwater development would allow for a better understanding of what groundwater irrigation means for the future. A recent study of the High Plains and the Ogallala Aquifer provides some of the information needed to address questions important to the future use of fossil water (8).

One possible strategy is to pump as fast as privately profitable until the water is too costly to pump and then appeal to society to pay for massive water transfers into the region. Although such a strategy is probably best for the region, it is not likely to be preferable for society. A better strategy might be a relocation subsidy and training program to help move people out of areas with excessive aquifer draw-down, such as the High Plains of Texas. With adequate planning, the option of slowing down pumping or developing alternative industry with lower water requirements remains open. This could be achieved by direct regulation or by a water tax equal to user costs. Society should decide whether or not the rates of groundwater use are too high before another technological change occurs or the farm economy prompts another rapid expansion of groundwater irrigation.

A related problem is well interference. This can occur either in areas with fossil water or in areas with good recharge rates. Administrative procedures and incentives pertaining to well interference should be designed to encourage private parties to negotiate prior to irrigation and minimize the chances of costly litigation (14). This may be a case where most of the competition is among a few local parties. Consequently, administrative or legal arrangements that make bargaining as easy as possible are probably the best solution.

Through its water laws, society has given a high priority to agricultural uses of water. While there is considerable institutional resistance to change, municipal and domestic uses generally have priority over agricultural uses. In some cases, the preference rule allows priority users to condemn the water of nonpriority users in times of scarcity in order to satisfy the priority users. Agricultural uses ranked high in the past because a higher percentage of the populace lived on farms and the economy was heavily dependent on agriculture. This priority is changing even in the North Central Region. The number of people living on farms is now quite small, and the relative importance of agriculture has declined. Consequently, in the future the North Central states may well place industry, energy production, and recreation in the second priority and relegate agriculture to third priority.

The returns per acre-foot indicate such a priority. Yet laws and administrative rules are slow to change. Thus, we should be concerned with changing these institutions so that they represent society's priorities and see that they do not restrict transfers to higher valued uses.

Competition from Industrial and Municipal Uses

Factories and homes use sizeable quantities of water in the large urban areas of the North Central Region. Water is used for cooling, human consumption, waste dilution, washing, lawn watering, and many other uses. Municipal and industrial uses are often quite competitive. For example, the demand for clean drinking water competes directly with waste dilution along a river or in lakes.

Chicago in the 1800s was a city that faced the trade-off between dumping wastewater into Lake Michigan and drawing drinking water from the same source. The importance of this trade-off was highlighted by the 1885 typhoid and cholera epidemic, which claimed 12 percent of the city's population (3). Chicago's solution to the problem was to reverse the flow of the Chicago River so that it emptied into the Illinois River rather than into Lake Michigan. Chicago's waste now flows down the Illinois River to the Mississippi River rather than polluting Lake Michigan. This eliminated the competition between waste dilution and water for human consumption, but created a new set of competitive situations.

The diversion lowered the level of the Great Lakes and created competition between water users near the Great Lakes and along the Illinois River. The lower lake levels meant potential problems for lake transportation and reduced electric power production. Chicago's diversion of 3,200 cubic feet of water per second from Lake Michigan lowered the Great Lakes by almost 3 inches. During periods of low lake levels, this reduced the clearance in a number of harbors, which, in turn, reduced the weight of cargo that vessels could carry into the harbors. The estimated increase in annual shipping costs during the 1960s caused by this reduced capacity was from $84,000 to $106,000 (3).

Reduced lake levels also reduced hydroelectric power generation on the Niagara and St. Lawrence Rivers and meant facilities were not used to capacity. The estimated loss in New York power production in 1966 was $3 million. However, in recent years higher lake levels have largely offset the loss; plants are running at or near capacity. In fact, the diversion may benefit some lakeshore areas by reducing erosion.

On the Illinois River, the diversion not only diluted waste loads but also increased navigation and electrical generation. The estimated annual navigation benefits were $150 million in 1981. Increased hydropower benefits fell in the $150,000 to $300,000 range. However, on the negative side, downstream flooding increased on the river, and recreational potential was reduced.

Thus, the potential for competitive uses of water for industrial and municipal uses in metropolitan areas is real and likely to increase during future drought periods. The cost of eliminating such conflicts could be substantial. Limits on lawn watering during dry periods may be inadequate in the future. Federal and state governments have already spent billions of dollars on municipal waste treatment plants to reduce the waste loads in rivers and streams. Even these expenditures have not had the impact on water quality that many had hoped for, partly because the money was spent on capital-intensive municipal waste treatment plants.

Another area of growing concern in the competition between the demand for clean water and waste disposal is groundwater pollution. Nitrate pollution is particularly critical in agricultural areas, especially in areas with sandy soils, where most irrigation takes place. Nitrate levels exceed safe limits in thousands of shallow private wells in the North Central Region. The primary sources of pollution appear to be livestock feedlots and household septic systems (7).

A closely related problem is the pollution potential from poorly cased and abandoned wells. Abandoned wells should be sealed to prevent contamination from the surface or mixing of water between aquifers. This is becoming an increasingly serious problem with the expansion in numbers of private wells.

Groundwater movement is extremely slow when compared with surface water flow. Once a pollutant is introduced into an aquifer, it takes a long time for it to be detected and flushed out. Detection of the pollution source is also difficult because the pollutant will move slowly through the aquifer. Thus, the effects of groundwater pollution are likely to be long lasting.

Another pressing concern is the danger of water pollution from landfills. Also, a growing number of manufacturing sites with chemical dumps threaten or already have polluted groundwater. Several states are dealing with this problem through regulation and establishment of hazardous waste disposal sites. It may be difficult to control unauthorized dumping unless public disposal sites are easily accessible and inexpensive to the disposer. Otherwise, unauthorized dumping will continue to be the least-cost, private alternative means of disposal. An alternative could be to raise the costs of unauthorized dumping by levying high fines on violators of pollution regulations and by more closely supervising firms that produce hazardous waste.

Another step would be to better coordinate existing programs and overlapping jurisdictions dealing with water quality. In Minnesota, for example, water quality activities involve 30 separate permitting authorities, nonpermitting regulatory activities, and monitoring and study programs (15). Thus, because of overlapping and capital-intensive efforts, water quality improvements have been costly and not very effective.

Water and Energy Production

Virtually every phase of energy production involves water, giving rise to issues related to water availability and quality. In the North Central Region currently, coal mining and electric power generation (from steam, hydroelectric, and nuclear plants) are the most significant energy-related activities. In the future, coal gasification and slurry pipelines could also create significant demands for water resources in the region.

Coal Mining

Coal, the largest fossil fuel resource in the United States, is found in significant deposits in the North Central Region, primarily in North Dakota, Illinois, and Ohio. (North Dakota produces lignite coal, the other states in the region primarily bituminous coal.) Both surface and underground mines exist in the region.

Mining and initial coal processing are the least water-consumptive phases in the use of coal. Neither surface mining nor underground mining requires large quantities of water. Coal from underground mines generally must be washed before use; consequently, water use is greater for underground mining (6). Water requirements for mining, regardless of type of operation, are quite small. Furthermore, mining does not pose a threat to other water users in the region by competing for the available supply. On the other hand, in certain local areas the quality of water supplies (particularly groundwater sources) may be adversely affected by acid mine drainage or salt percolation, which could affect downstream users.

Electric Power Generation

The largest use of water in the energy sector is for cooling electric power plants. In 1975, there were 297 coal-fired electric generation plants of over 100 megawatts in the United States, and 146 of those plants were located in the North Central Region (Table 2). Of particular importance in considering water use and quality dimensions of coal-fired power plants is the technology used for cooling. The most important function

Table 2. *Number of coal-fired electric power plants, capacity, and water requirements, North Central Region, 1975* (2).

State	Number of Plants	Installed Capacity (megawatts)	Water Requirements (cubic feet/second)			Capacity Using Cooling (megawatts)
			Withdrawal	Discharge	Consumption	
Illinois	22	16,460	8,577.3	8,426.1	151.2	16,196
Indiana	20	12,801	11,363.7	11,192.9	170.8	12,713
Iowa	11	2,689	2,135.0	2,100.5	34.5	2,683
Kansas	6	2,481	1,544.1	1,507.4	36.7	2,481
Michigan	16	12,094	10,051.7	9,948.0	103.7	11,935
Minnesota	9	2,781	1,871.6	1,846.7	24.9	2,776
Missouri	14	8,563	4,320.2	4,276.9	43.3	7,424
Nebraska	3	987	900.3	892.7	7.6	986
North Dakota	4	1,208	685.9	677.2	8.7	1,208
Ohio	28	22,106	18,895.3	18,441.7	453.6	21,350
South Dakota	1	455	3.5	0.0	3.5	455
Wisconsin	12	4,948	4,640.7	4,575.5	65.2	4,894
Total	146	87,573	64,989.3	63,885.6	1,103.7	85,101

of water in such plants is to dissipate waste heat. This can be accomplished through wet cooling or dry cooling techniques or a combination of the two. Wet cooling processes (which include once-through cooling, cooling ponds, and wet towers) dissipate heat through circulating water. Dry cooling, on the other hand, dissipates heat through air. The technology used depends upon the availability of water, environmental impacts, and cost of investment. Dry cooling uses no water, but is the most costly. Wet cooling requires varying amounts of water depending upon the process used (Table 3). Once-through cooling draws water from a stream or lake through the system and then discharges the water back to its source. This technique results in the greatest withdrawal of water, but is lowest in actual consumption. It requires abundant water supplies and no legal limits on thermal pollution.

A wet-cooling tower is a closed system that requires lower water withdrawals, but has a higher consumption rate than once-through cooling. Such a system is used when water must be obtained from groundwater sources or when the discharge of effluent must be avoided.

Cooling ponds consume even more water as heat is dissipated through surface evaporation. The capacity of a power plant dictates the size of the cooling pond. From one to two surface acres are required per megawatt of capacity.

Coal-fired generating capacity, as measured in megawatts, dominates nuclear capacity by seven to one in the United States (2). Nuclear power generation, however, cannot be ignored, because on a per megawatt basis it requires 40 to 50 percent more water than fossil-fired plants (25). Of the water resources regions within the North Central Region, the Upper Mississippi and Great Lakes Regions have the most on-line nuclear capacity and the greatest potential for future development (21).

Hydroelectric power historically has been an important power source in the United States, but because few sites remain for development, it will not grow in importance. In 1974, 21 percent of the installed hydroelectric capacity in the United States was accounted for by plants in water re-

Table 3. Cooling technology and water consumption for coal-fired power plants (2).

Technology	Amount of Water Consumed (cubic feet/second) Per 1,000 Megawatt Capacity
Dry cooling tower	0.0
Once-through cooling—fresh	7.8
Once-through cooling—saline	12.2
Wet cooling tower	12.5
Cooling ponds	15.4
Combination	18.1

sources regions within or adjacent to the North Central Region (21). Hydropower offers several advantages over other methods of producing electric power. It is often compatible with the provision of recreation, water supply, and flood control. Furthermore, power is produced without consuming fuel and without polluting air or water.

Coal Gasification

Production of synthetic fuels in the United States has never received the prominence it has in other countries, such as South Africa (16). However, a coal gasification plant with the capacity to produce 250 million cubic feet of gas per day is under construction in western North Dakota and will be completed in 1984 or 1985. Estimates of water requirements for such plants vary greatly, from 10,000 acre-feet to 45,000 acre-feet of water per 250 million cubic feet of gas produced per day (26). Because most of the water is needed for cooling purposes, dry cooling could significantly reduce water requirements.

Coal Slurry Pipelines

Two coal slurry pipelines now operate in the United States. One carries coal from Arizona to Nevada, the other is a 108-mile pipeline in Ohio. A number of other pipelines have been proposed, and in some cases planning has advanced to the stage of attempting to secure water permits or easements. Recently, Energy Transportation Systems Inc. entered into an agreement with the State of South Dakota to purchase water from the Missouri River to use in a slurry connecting Wyoming and Arkansas.

Coal slurry is relatively water intensive: One ton of water is required to move one ton of coal (25). On the other hand, it can provide a low-cost alternative to other modes of transportation (primarily rail), especially over long distances. The major criticism of coal slurry pipelines is that water is exported out of the basin and is not returned. On the other hand, water is not consumed but reused at its destination.

Energy Production Issues

Energy development is frequently seen as a threat to water resources, both in terms of quantity and quality. In the North Central Region, water quality issues are much more significant than those related to quantity.

Water Availability

Water supplies are generally adequate for energy production across the North Central Region, so competition between energy and other uses will be a problem only in a few local areas. A review of trends in water withdrawals and consumption by use indicates that total withdrawals for all uses will decline by the year 2000, but consumption will increase (*22, 23, 24*). While total withdrawals will decline, the percentage of withdrawals for steam electric generation will increase, along with the percentage of consumption. This could lead to conflicts in use, particularly during periods of low stream and river flows.

In western North Dakota, where substantial energy development is occurring along the main stem of the Missouri River, water can be obtained from either the river or Lake Sakakawea. Therefore, the physical availability of water will not be a problem.

Metropolitan areas within the region that have potential water supply problems that could affect steam electric generation are Minneapolis-St. Paul, Peoria-Pekin, St. Louis, and Chicago (*24*). In dry periods, locally high withdrawal and consumptive use could create water supply and instream problems.

The Great Lakes generally provide an adequate supply of water to support aquatic and human activities in states contiguous to the lakes. However, lake levels do fluctuate, and during periods of low levels, it may be difficult to maintain full power production from hydroelectric power plants.

Water Quality

The most important issue surrounding the use of water for energy development and production of electric power is water quality. Depending upon the type of energy-related activity, the issue may be groundwater or surface water pollution, or it may be disposal of low-quality waters in areas far removed from their origin (*13*).

Where coal mining is the principal form of energy development, the water quality issue is acid mine drainage, which may occur with either underground mines or strip mines. Acid mine drainage is particularly significant in the Ohio region and southern portions of the upper Mississippi Water Resources Region.

Two-thirds of the nation's acid mine drainage problems occur in the Ohio Water Resources Region. In 1978, it was estimated that it would cost $350 to $450 million to prevent acid drainage from abandoned deep mines in the region (*21*). The cost to control water pollution resulting from mine drainage in Illinois was an estimated $346 million (*24*). Industrial uses, municipal uses, navigation, and recreation are all affected by mine drainage. Because many industrial firms must treat the water before they use it, costs of production, and presumably, final product prices, are increased. To make water drinkable, municipalities frequently require treatment procedures they would not otherwise need. The life-span of some water-based transportation equipment is shortened as a result of acid mine drainage. Because the acidity causes fish and aquatic life kills, recreation is also affected.

Thermal pollution is caused by the discharge of heated water from a power plant into a lake, river, or stream. These discharges may cause drastic changes in aquatic life downstream. The discharge of heated water has both positive and negative effects. It is beneficial to warm water fish species, but it also promotes growth of algae, which may restrict some recreational activities and create the need for water treatment downstream by municipal and industrial users. Research on discharges of heated water into Lake Michigan has shown that most of the waste heat is discharged into the beach water zone, the area of the lake most important for human use (21).

Water for Outdoor Recreation

Many forms of outdoor recreation are water dependent or water oriented. Of those that are water dependent, fishing, sailing, rafting, swimming, and canoeing are the most popular. In fact, sailing is expected to increase by 267 percent from 1977 to 2020.

Water-based outdoor recreation depends upon adequate in-stream flows and flatwater recreation areas. Water quality is extremely important for contact activities, such as swimming or skiing, but much less important for other activities, such as boating or sightseeing. It has been argued that about 60 percent of the $10.1 billion in annual damages attributable to all water pollution in the United States is associated with the loss of water-based recreational opportunities (5).

Recreational water use has only recently been recognized as a national resource worthy of development. Meanwhile, the demand has continued to exceed the supply in many areas. Recreation is included as a beneficial product of publically sponsored water projects, such as dams for flood control or hydropower (18). The courts have sometimes required that water be withdrawn from streams to establish a beneficial use, thus making it difficult to establish water rights for recreational pursuits.

Water is used in a variety of ways for recreation. The requirements in terms of quality and quantity vary among uses. Recreational users of water may be quick to notice changes in water quantity. Marginal changes in water quality are less obvious, especially for noncontact activities. In the Great Lakes, for example, the most serious types of water pollution are toxic substances carried through the food chain and accumulated in fish and nutrient enrichment, which results in accelerated eutrophication and unpleasant effects on the appearance, odor, and taste of water. These water quality problems are much less apparent, unless they are very serious, than are changes in lake or stream levels.

The North Central Region, with 28 percent of the U.S. population, has 34 percent of the licensed fishermen (9.5 million licenses in 1980). In addition, 35 percent of the pleasure boats (more than 3 million) are registered in North Central States. The U.S. population is expected to increase from 226 million in 1980 to 283 million in the year 2000. If the proportion of fishermen remains relatively constant, the absolute number of fishermen will increase considerably. Fishing, however, is but one of many water sports.

The intensity of fishing in the North Central Region gives some indication of the supply of water-based recreational opportunities. Minnesota is the "land of 10,000 lakes," and Michigan has 11,000 inland lakes and 3,288 miles of shoreline on the Great Lakes. The Ohio River, one of the major river systems in the region, is a 981-mile long recreational "lake" with about 400 small boat facilities. Federally developed water access points attract more than 2 million recreational visitors per year. Water-based recreation in the upper Mississippi Basin accounted for 185 million recreation days in 1975 and is expected to provide 284 million recreation days in the year 2000 (*24*). Though this is just a sample of the water-based recreation existing in the North Central Region, these figures illustrate the diversity and pervasiveness of the region's water-related recreation.

Water-based recreation often competes for water resources on many fronts because of its diversity and pervasiveness. There may be competition among recreators, such as crowding at public beaches or at good "fishin' holes." Power boaters often compete with sail boaters or fishermen. Additionally, waterfowl hunters may contest canoeists for stretches of rivers.

Water quantity issues may be significant when water used in recreation is withdrawn by extensive uses, such as agriculture or flood control storage. The quantity of water available for outdoor recreation in the North Central Region will not be a constraint for some time, except in areas with high population concentrations or excessive withdrawals.

Recreational opportunities may also be subject to impacts by draw downs of groundwater aquifers. This could occur where wetlands or small streams are maintained through high groundwater tables, as in the Sandhills of Nebraska.

The primary availability issue stems from the fact that many of the region's recreational resources are not readily accessible to the region's residents, particularly those living in the metropolitan areas. Chicago, Detroit, and Cleveland, for instance, contain about one-half of the Great Lakes Region's population but only 4 percent of the recreational land and water supply (*19*).

There may be both competition and complementarity among all types of water-based recreation and other users of water. For example, flood control structures create slack water reservoirs that provide immense potential for fishing and boating. Dams may change the nature of the resource, such as in North Dakota, where the Missouri River was changed into a large reservoir supporting a popular salmon fishery in addition to the native species.

Competition may be easier to identify than complementarity. Barge traffic may interfere with recreational use or compete for space on major rivers in the region. Operation of a dam for flood control may cause water levels to fluctuate more drastically than is desired by owners of recreational property on the reservoir shoreline. Clearly, dumping of municipal or industrial wastes reduces the downstream opportunities for water-based recreation. However, dumping of industrial cooling water frequently creates a fishing hotspot or an open-water pool for wintering waterfowl, thereby enhancing recreational opportunities.

The degree of competition varies with the intensity of competing use. For example, extremely high levels of municipal waste discharges would certainly affect swimming or fishing, but moderate discharges may have little impact. Blanket statements regarding competitive uses should not ignore the intensity of use effect or the amelioration brought about by proper timing and management of discharges.

Water quality issues are more pronounced, yet less subject to measurement than are availability issues. Water pollution produced by soil erosion and other nonpoint sources may impair the attractiveness of water-based recreation. However, identification of the competing party is difficult. Another example of indirect competition is in the case of acid rain. Reduced pH in many lakes has caused a deterioration in certain recreational activities. The competing use is atmospheric disposal of nitrates and sulfates, pollutants that ultimately reach water bodies. One use clearly affects the other, but the path of competition is difficult to follow.

Most of the competition resulting from water-based recreational activity results from the low price or lack of a charge for most water-based recreation. Such a pricing policy encourages overuse and crowding. Institutional arrangements, such as high user fees and local financing, have been suggested as mechanisms for internalizing the crowding externalities. The question of property rights pervades the issues. More explicitly, who owns the right to what quality of water?

Navigation as Competition

Within the North Central Region lie the headwaters and a large portion of the longest river in the world, four of the five Great Lakes, and numerous navigable river systems. This abundance of navigable waters has spawned an extensive network of water-borne commerce reaching throughout this huge grain-producing region.

The major rivers in the region—the Mississippi, Ohio, Illinois, and Missouri—each support water-borne commerce. Barges carry about 10 percent of all ton-miles of freight in the United States. More than $1 billion per year has been spent on the federal rivers and harbors program (10).

Inland waterway traffic is expected to nearly double between now and 2000—from about 500 to 900 million short-tons per year. Mississippi River traffic between Minneapolis and New Orleans amounts to more than 300 million tons per year. The Ohio River flows 981 miles—from Pittsburgh to Cairo, Illinois. The entire river has been improved by construction of locks and dams. U.S. Army Corps of Engineers' projects have accounted for more than $1 billion of expenditures on the Ohio River (10). About 140 million tons of water-borne freight are handled each year on the Ohio, two-thirds of which are bulk forms of energy, coal, crude oil, and petroleum products.

The ports of the Great Lakes are part of a global network, with ships from many nations docking in Duluth, Detroit, and other cities. Lake Michigan alone accounted for more than 114 million short-tons of cargo in 1978, of which about 12 percent was foreign bound. The Soo Locks between Lakes Superior and Huron handle more cargo every year than does the Panama Canal.

Principal navigational uses of water are to provide a means and a substance on which to float barges and other carriers. Though primarily nonconsumptive, navigation accounts for considerable movement of water through the operation of locks and maintenance of in-stream flows. Requirements are a navigable channel maintained free of obstructions. Channel depth and lock capacity are the controlling dimensions limiting the amount of cargo per unit as well as unit draft.

Some of the operations necessary to accommodate navigation in the North Central Region affect both water quantity and quality available for other uses. Therefore, navigation competes for available water resources. As with other uses of water, competition can be among users for the same use, or among uses or geographic areas. The size and shape of channels and capacity of locks are constraints that create competition at certain bottlenecks within navigational uses. Tugs pushing barges up the Mississippi may be queued at locks awaiting their respective turns.

Dredging to maintain 9-foot channels may cause local water quality problems and reduce recreational opportunities. Maintenance of minimum stream flows to accommodate navigation, especially in periods of low flows, can take water away from out-stream uses, such as irrigation. In-stream flow maintenance is currently at issue in the Missouri River system. Downstream users want adequate flows to support navigation, while upstream users want to withdraw water for agriculture and energy development. South Dakota's recent sale of water rights from the Oahe Reservoir for a coal slurry has heightened this conflict among water users and has renewed interest in interbasin transfers. Water necessary for lock operation may be used for power generation or municipal supplies. Surges from peak power generation may interfere with navigation (*10*). However, peak power demand may not occur when water requirements for navigation are greatest.

The contemporary issue of concern regarding navigation in the North Central Region is that of users' fees. Currently users are paying a five-cent tax per gallon of fuel to support operation and maintenance of navigational works. This tax is scheduled to increase in the near future. The waterway user's fee creates a particularly complex situation, because it is difficult to allocate costs among users. If barges pay a user's charge, then perhaps recreators should also pay, but what cost is imposed upon the system by recreational users?

Other Kinds of Competition

Five commonly cited types of water use and some typical areas of competition for water have been discussed. There are several uses that do not fit neatly into these groupings and do not as clearly result in competition. Acid rain, resulting from air pollution, produces water pollution that creates competition between direct water users and air polluters who indirectly use water. There are also competitive situations that are more difficult to describe and quantify. For example, wetlands drainage or commercial development of riverine habitat may contribute to the demise of threatened or endangered wildlife species. Society proclaims it values these species and desires that their habitat be maintained, but private development may prevail. Thus, competition exists through intricate interrelationships, as well as through those more clearly understood.

Flooding—both upstream and downstream—is a significant issue in all areas of the North Central Region. Flood control structures compete with navigation, complement hydropower generation, and create and destroy recreational opportunities. Upstream drainage and channel improvements can result in increased flooding downstream.

Legal and institutional constraints to solving water problems and use conflicts abound in the region and are especially prevalent in the Northern Plains States. The Winters Doctrine claims of the Native Americans and the Reserved Rights Doctrine of the federal government clash with existing patterns and practices in many states (*12*). Neither the Indians nor the federal government have fully documented or identified their claims to waters. This presents considerable uncertainty when planning future water resource use.

A final issue confounding the competition for water from society's perspective is the distribution of income. Welfare implications surround every water management decision; there are both winners and losers. In making public decisions regarding water allocation between competing uses, the distribution of income both before and after the decision must be considered. In most cases there are several uses and parties involved, making welfare implications an extremely complex consideration.

A Future of Increasing Competition

In summary, competition for water exists in a variety of forms and under varying conditions. Timing, location, quantity, and quality each affect the degree of competition. Competition exists among users, between uses, among geographic areas, and across time. It can be direct and obvious, as in the case of barges versus fishermen or among irrigators, or it can be indirect and subtle, as in the case of acid rain.

Each of five major water uses—irrigation, municipal and industrial uses, energy development, recreational use, and navigation—in the North Central Region exhibits degrees of competition and complementarity. The intensity of the relationship, while depending upon a number of factors, is extremely sensitive to climatic cycles. Drought years heighten the competition for water between irrigators and navigational users. Wet years heighten the competition between drainage enterprises and downstream floodplain residents. In the first instance the competition is expressed in competing claims for receiving water, but in the second the claims involve disposal of excessive water.

In conclusion, competition for water, the capricious resource, exists from one-dimensional intrause (among irrigators) to multidimensional interuse (among irrigators, recreators, navigation, etc.) and across time. In normal years in the North Central Region there is little competition, with the exception of localized problems. However, in other years—too dry or too wet—more competitive situations arise. With increasing water demands that are expected in the future, will come increasing competition for water in the North Central Region.

REFERENCES

1. Boris, Constance M., and John V. Krutilla. 1980. *Water rights and energy development in the Yellowstone River Basin: An integrated analysis.* The Johns Hopkins University Press, Baltimore, Maryland.
2. Dalsted, Norman L., and John W. Green. 1981. *Water use by coal-fired power plants. 1975.* Staff Report No. AGESS810326. Economics and Statistics Service, U.S. Department of Agriculture, Washington, D.C.
3. Easter, K. William, and John J. Waelti. 1980. *The application of project analysis to natural resource decisions.* WRRC Bulletin 103. Water Resource Research Center, University of Minnesota, St. Paul.
4. Frederick, Kenneth D., and James C. Hanson. 1982. *Water for western agriculture.* Resources for the Future, Washington, D.C.

5. Freeman, A. Myrick, Robert H. Haveman, and Allen V. Kneese. 1973. *The economics of environmental policy*. John Wiley and Sons, Inc., New York, New York.
6. Gray, S. Lee, Edward W. Sparling, and Norman K. Whittlesey. 1979. *Water for energy development in the Northern Great Plains and Rocky Mountain regions*. ESCS Staff Report NRED 80-3. Economics, Statistics, and Cooperatives Service, U.S. Department of Agriculture, Washington, D.C.
7. Great Lakes Basin Commission. 1975. *Great Lakes basin framework study (Appendix 6: Water Supply—Municipal, Industrial, and Rural)*. Ann Arbor, Michigan.
8. High Plains Associates. 1982. *Six state High Plains Ogallala Aquifer Regional Resource Study*. U.S. Department of Commerce, Washington, D.C.
9. Howe, Charles W. 1979. *Natural resource economics issues, analysis and policy*. John Wiley and Sons, New York, New York.
10. Howe, Charles W., Joseph L. Carroll, Arthur Hurter, Jr., William J. Leininger, Steven G. Ramsey, Nancy L. Schwarty, Eugene Silberberg, and Robert M. Steinberg. 1969. *Inland waterway transportation: Studies in public and private management and investment decisions*. The Johns Hopkins Press, Baltimore, Maryland.
11. Howe, Charles W., and K. William Easter. 1971. *Interbasin transfers of water economic issues and impacts*. The Johns Hopkins Press, Baltimore, Maryland.
12. Kneese, Allen V., and F. Lee Brown. 1981. *The Southwest under stress. National resource development issues in a regional setting*. The Johns Hopkins University Press, Baltimore, Maryland.
13. Larson, William E., Leo M. Walsh, B. A. Stewart, and Dan H. Boelter. 1981. *Soil and water resources: Research priorities for the nation*. Soil Science Society of America, Inc., Madison, Wisconsin.
14. Lotterman, Edward D., and John H. Waelti. 1980. *Efficiency and equity of alternative well interference policies in semi-arid regions*. Staff Paper P80-19. Department of Agricultural and Applied Economics, University of Minnesota, St. Paul.
15. Minnesota Water Planning Board. 1979. *Towards efficient allocation and management: A strategy to preserve and protect water and related land resources*. St. Paul, Minnesota.
16. Stobaugh, Robert, and Daniel Yergin. 1979. *Energy future*. Random House, New York, New York.
17. Supalla, Raymond J., and Robert R. Lansford. 1982. *Resource policy implications of the High Plains Ogallala Aquifer study*. Department of Agricultural Economics, University of Nebraska, Lincoln.
18. U.S. Army Corps of Engineers. 1981. *78-79 recreation statistics*. Washington, D.C.
19. U.S. Army Corps of Engineers. 1981. *Water resources development in Michigan*. U.S. Government Printing Office, Washington, D.C.
20. U.S. Department of Agriculture, Soil Conservation Service. 1977. *Basic statistics: 1977 National Resources Inventory (NRI)* (Revised February, 1980). Washington, D.C.
21. U.S. Water Resources Council. 1974. *Water for energy self-sufficiency*. Washington, D.C.
22. U.S. Water Resources Council. 1978. *The nation's water resources, 1975-2000. Volume 4: Great Lakes region*. Washington, D.C.
23. U.S. Water Resources Council. 1978. *The nation's water resources, 1975-2000. Volume 4: Ohio region*. Washington, D.C.
24. U.S. Water Resources Council. 1978. *The nation's water resources, 1975-2000. Volume 4: Upper Mississippi region*. Washington, D.C.
25. Weatherford, Gary, editor. 1982. *Acquiring water for energy: Institutional aspects*. Water Resources Publications, Littleton, Colorado.
26. Western States Water Council. 1974. *Western states water requirements for energy development to 1990*. Salt lake City, Utah.

III
Water Research Needs
and Potentials

9

Water Resources Research: Potential Contributions by Plant and Biological Scientists

M. B. Kirkham

Water stress affects photosynthesis and other plant physiological processes probably more than all other environmental factors combined (*50*). Even in the humid North Central States, water almost always limits growth at one stage in a crop's life cycle (*42, 50*). Therefore, water resources research in the plant sciences is important. Research priorities have been listed (*9, 10, 17-19, 26, 27, 31, 40, 43, 45*).

Water Conservation

Comparison of Plant Water Uses

Research related to water conservation is the most important need in Kansas (*17*) and a major need in other North Central States (*26, 27, 40*). Groundwater from the central Ogallala formation is the source of irrigation water for the Great Plains. In many areas, including Kansas, withdrawal from the aquifer exceeds recharge. Farmers are abandoning land because they no longer can afford to pump water from the deepening groundwater source. If this water is not conserved, farmers in the Great Plains will have to revert to dryland techniques (*6, 49*), resulting in lower yields and economic disorder. Irrigation from the aquifer will be short-lived unless conservation measures are implemented (*5*). One way to conserve water is to identify crops that use less water.

For at least 70 years, it has been known that such plants as corn (*Zea mays* L.), sorghum (*Sorghum bicolor* Moench.), and millet (*Pennisetum americanum* Auth.) have lower water requirements than such plants as dry beans (*Phaseolus vulgaris* L.), soybeans (*Glycine max* Merr.), and sunflowers (*Helianthus annuus* L.) (*3, 4*).[1] In the past 20 years, researchers have determined that corn, sorghum, and millet have the C-4 type of photosynthesis in which the primary carbon dioxide-fixation products have four-carbon atoms (oxaloacetic acid, aspartic acid, malic acid). Dry beans, soybeans, and sunflowers have the C-3 type of photosynthesis in which the first product formed from carbon dioxide is three-phosphoglyceric acid (*36*). In C-4 species, because leaf photorespiration (loss of carbon dioxide in the light) is difficult to detect, C-4 plants apparently recycle carbon dioxide within the leaf. This results in a higher rate of net photosynthesis and a lower transpiration ratio for C-4 plants than for C-3 plants (*2*). Recent research (*13*) confirms that C-4 plants have a lower water requirement than C-3 plants (Table 1).

There has been little research done to compare the internal water status of C-3 and C-4 plants. Most studies measure the amount of water a crop receives and the final grain yield. Researchers do not look at water *in the plant*. To understand why the water requirement of plants varies, it is necessary to determine the water relations of the plants themselves.

The classical plant-water equation (*23*) is

$$DPD = OP - TP$$

where DPD is the diffusion pressure deficit, OP is the osmotic pressure, and TP is the turgor pressure. The new terms for these factors are water potential, osmotic potential, and turgor potential, respectively (*32*).

Table 1 gives the water potential, osmotic potential, and turgor potential of the six crops mentioned above. Corn, the crop with the lowest water potential, also has the lowest water requirement. Soybeans and pinto beans have relatively high water requirements and the highest water potential. Corn has a lower turgor potential than soybeans and a lower osmotic potential than soybeans or pinto beans. The higher water potential and turgor potential of pinto beans and soybeans, compared to corn, indicate that they lose more water than corn. The water requirement values, obtained independently (Table 1), also show that pinto beans and soybeans use more water for the same grain yield than corn.

It is important to understand how the internal water balance of plants affects their water requirements. Contributions that plant scientists might make in this area include the following:

1. Do C-4 plants, in general, have a lower water potential, osmotic potential, and turgor potential than C-3 plants?

2. Do low potentials enable plants to have low water requirements?

---

[1]Water requirement is "the ratio of the weight of water absorbed by a plant during its growth to the weight of dry matter produced" (*3*). The water requirement has also been called the transpiration ratio. See reference *42* for a discussion of definitions.

3. How is plant morphology related to plant water balance? (Corn leaves, with a low water requirement, are more fibrous and stiff than bean leaves, with a high water requirement.)

4. How does the internal water balance and water requirement of plants vary when plants are grown under dry conditions compared with when the same plants are grown under well-watered conditions? Values in table 1 apply to well-watered plants. Would the same relative values be obtained for plants exposed to drought?

Answers to these questions would permit an understanding of how partitioning of water in a plant affects its water requirement. This would be of practical value because plants identified as having low water requirements could be planted to conserve water.

Studies of Plant Roots

Plant roots are the organs through which water is absorbed. Therefore, research on plant roots is of primary importance in water conservation studies. Such research will take much time, labor, and money because measurement of root growth and extraction of roots from the soil are time-consuming and tedious. For these reasons, there have been few experiments done with roots, compared to those done with shoots. There are four specific areas in which plant scientists could contribute through plant root studies.

Roots for Reduced Tillage. Reduced tillage is an area of intensive study now because it reduces energy costs. It also has the added benefit

Table 1. *Internal water status and water requirement of six field crops grown at Manhattan, Kansas, in the summer of 1981. Values for potentials are the average of 12 measurements taken weekly between June 25 and September 17.*

Crop	Water Potential	Osmotic Potential	Turgor Potential	Water Requirement $\left(\dfrac{\text{kg water/ha}}{\text{kg grain/ha}}\right)$
		bars		
Corn	− 19.4a*	− 24.7a	5.3b	606†
Millet	− 18.9a	− 25.6a	6.7ab	2540‡
Sorghum	− 13.1bc	− 20.1b	7.0ab	704
Sunflower	− 14.3b	− 19.9b	5.6ab	2100
Soybean	− 11.2c	− 18.3b	7.1a	1580
Pinto bean	− 8.7d	− 14.9c	6.2ab	1630

*In each column, measurements followed by the same letter are not significantly different at the 0.05 level, according to Duncan's new multiple-range test.
†Water requirement calculated from data provided by Hattendorf (*13*), who describes how the plants were grown.
‡Water requirement for millet is probably too high because birds ate much of the grain before harvest.

of conserving soil water. Mulch on the soil surface retards evaporation. Scientists cite soil compaction as a major problem with reduced tillage, and one reason why yields with minimum tillage are almost invariably less, throughout the world, than with normal plowing (16). With little plowing, soil is compacted, and rooting in soils with great resistance is less than in plowed soils. Is it possible to find root systems of crops that are better able to penetrate soils of high resistance than those of other crops and thereby use soil water that otherwise would be unavailable for growth? At the Kansas State University Evapotranspiration Laboratory, scientists are studying alfalfa (*Medicago sativa* L.), corn, and soybeans in a soil with a hard pan to see if the roots vary in their ability to grow through the soil. Data from the first summer's work (1982) showed that corn penetrated the high-strength soil as well as alfalfa did (unpublished data). These results were surprising because scientists generally believe that alfalfa has a deeper root system than corn, which often has its roots confined to the upper layers of the soil (11, 30).

Potential contributions that plant scientists could make in the area of rooting under reduced tillage include the following:

1. Can root systems of different species and varieties be identified that grow better under reduced tillage than other root systems?

2. Can certain crop rotations be identified that result in higher yields under reduced tillage compared with other rotations? For example, does a deep-rooting perennial legume, such as alfalfa, improve soil structure to a degree that justifies a regular rotation of alfalfa with annual crops, such as corn and soybeans?

3. How do the four soil physical properties—water, temperature, air, and bulk density—affect root growth of various species and varieties grown under reduced tillage?

Answers to these questions perhaps would eliminate the yield discrepancy between reduced tillage and regular plowing.

Comparison of Roots of Varieties Differing in Height. Semidwarf wheats (*Triticum aestivum* L. em. Thell.) are of great economic importance in Kansas, accounting for over half of the wheat planted. There is concern in the state and in other Great Plains States, however, that semidwarf varieties may be more susceptible to drought than standard-height varieties because root and shoot development are interdependent. Roots and shoots are considered mirror images of each other (22). Roots depend upon the shoots for photosynthates, and shoots depend upon the roots for water, nutrients, and manufacture of such hormones as cytokinins. But research shows that roots of tall and short varieties of winter wheat appear to have the same amount of roots and grow at the same rate (15, 29, 33).

Specific questions that plant scientists could answer include:

1. How can a shorter shoot produce as many roots as a taller shoot?

2. Do tall varieties produce more roots than necessary? What is the minimum number of roots necessary for maximum yield?

3. Are the roots of tall and short varieties equally effective in absorbing water from the upper levels of the root zone as from the lower levels of the root zone?

4. Do tall or short plants have lower water requirements and, therefore, conserve more water?

Again, results from studies of roots of tall and short varieties of wheat would have immediate, practical value because farmers in Kansas want to know what type of wheat they should plant for maximum yield with the limited amount of soil water available.

Comparison of Roots of Drought-Resistant and Drought-Sensitive Plants. Different varieties of the same species vary in their ability to yield under drought. But apparently no studies have compared the root systems of drought-sensitive and drought-resistant plants under field conditions. It is important to know both the extent of roots and the amount of water taken up by roots of plants varying in drought resistance. Specifically, the following questions need to be answered:

1. Do drought-resistant plants have a lower water requirement than drought-sensitive plants?

2. Do drought-resistant plants have deeper penetrating roots than drought-sensitive plants? And, therefore, are they better able to extract soil water at lower depths than drought-sensitive plants? It may be that, in areas with limited amounts of rain, such as Kansas, roots of resistant plants do not grow deeply, but spread widely in the surface of the soil, where moisture is replenished by summer showers (*48*).

3. What are the water absorption zones of drought-resistant and drought-sensitive plants? Are drought-resistant plants more effective in absorbing water from lower (or upper) zones than drought-sensitive plants?

Researchers need to determine the characteristics of the root systems of drought-resistant plants to help understand why drought-resistant plants are able to grow with limited amounts of water.

Studies of Roots with Mycorrhizae. Microorganisms usually have detrimental effects on the water balance of plants (*8*). Rusts, for example, increase the evapotranspiration rate of wheat and make wheat more susceptible to drought (*41*). Little research has been done to show possible beneficial effects of microorganisms. Mycorrhizae are symbiotic associations between fungi and the roots of higher plants (*34*). Hyphae of the fungi, which penetrate the soil and infect the roots, enlarge the volume of soil from which a plant can absorb water and nutrients. Uptake of phosphorus particularly is increased in the presence of mycorrhizal roots because phosphorus is relatively immobile in the soil and roots must be close to phosphorus to absorb it. It would seem reasonable, therefore, that mycorrhizal roots would be an advantage during drought.

There has been little work done with mycorrhizal roots exposed to dry conditions even though there are indications that plants with mycorrhizal roots might be more resistant to drought than plants without mycorrhizal roots (28, 35). Recent work at Kansas State University showed that under both well-watered and water-stressed conditions corn with mycorrhizae had a higher water potential, osmotic potential, and turgor potential than corn without mycorrhizae (Table 2) (unpublished data). Also, under both wet and dry conditions, plants without mycorrhizae grew less than plants with mycorrhizae. Therefore, under drought, both growth and the three potentials increased when mycorrhizae were present as opposed to when they were absent. The results for growth and plant-water status agree, as water potential and turgor potential both relate directly to growth (7, 12). The results raise the following questions:

1. Do mycorrhizae affect the water status of plants? Previous results with onions (*Allium cepa* L.) suggest that mycorrhizae do not change

Table 2. *Effect of mycorrhizae (*Glomus epigaeus*) and phosphorus fertilization on internal water status and growth of corn (*Zea mays*) in a greenhouse.* *

Treatment	Water Potential	Osmotic Potential	Turgor Potential	Shoot Dry Weight	Root Dry Weight
	———————— bars ————————			——— grams ———	
Drought-stressed Nonmycorrhizal with 20 ppm P added	−14.8a†	−14.8ab	0a	3.49c	0.59c
Mycorrhizal with no P added	−11.3b	−12.6bc	1.3b	7.18ab	1.29b
Mycorrhizal with 10 ppm P added	−8.3c	−10.3d	2.0c	6.93b	1.45b
Well-watered Nonmycorrhizal with 20 ppm P added	−12.9ab	−15.2a	2.3c	7.34ab	1.21b
Mycorrhizal with no P added	−11.6b	−13.0abc	1.4b	7.76ab	2.05a
Mycorrhizal with 10 ppm P added	−8.1c	−11.9cd	3.8d	8.06a	1.61ab

*Plants were grown in 30-centimeter diameter pots, one plant per pot, which had a Naron-Farnum sandy loam with 6 parts per million available phosphorus. The soil was watered daily (well watered) or watered only when the corn was wilted (drought stressed). The corn was planted on November 12, 1981, and harvested for dry-weight measurements eight weeks later. Potentials, obtained with thermocouple psychrometers, are the average of 18 values taken on weekdays at about 8:00 a.m. between December 14, 1981, and January 8, 1982.

†In each column, measurements followed by the same letter are not significantly different at the 0.05 level, according to Duncan's new multiple-range test.

plant-water relations (28). But the results in table 2 indicate that the mycorrhizae alter the internal water status of plants that they infect.

2. Would results similar to those in table 2 be observed for crops other than corn?

3. Do varieties of crop species vary in ease with which mycorrhizae can infect their roots?

Knowing that mycorrhizae increased a plant's adaptability to drought would permit use of them immediately. Spores of mycorrhizae could be added to a drought-prone soil. After germination of spores, mycorrhizae then could infect susceptible crops, which might be more drought-resistant than if they were not infected.

Comparison of Plant Temperatures

If plants are water-stressed, leaf stomata close, the leaf is not cooled by transpiration, and the canopy temperature rises (24). Even under well-watered conditions, certain crop varieties tend to keep their stomata more closed and have higher temperatures than other varieties. For example, under well-watered conditions, corn hybrids known from experience to be drought-tolerant in the field have an average seasonal canopy temperature that is about 0.5 °C higher than that of hybrids known to be drought-susceptible (unpublished data, Kansas State University). Differences in stomatal opening appear to be genetically determined and, therefore, heritable. Thus, measurement of leaf temperature appears to be a promising new method to isolate water-conserving crop varieties. Measurements are easy because new, lightweight, hand-held infrared thermometers are now available commercially. Consequently, hundreds of varieties can be evaluated in a short period of time.

Plant scientists need to screen varieties of a species for differences in canopy temperature. Specifically, the following questions need to be answered:

1. Can canopy temperature, as sensed by an infrared thermometer, be used to screen varieties for increased drought resistance?

2. Do varieties known to be drought-resistant have a higher temperature than varieties known to be drought-sensitive?

3. How does canopy temperature vary with environmental conditions? That is, do drought-resistant varieties have high temperatures both under dry and well-watered conditions? Do drought-resistant crop varieties have high temperatures both under low and high relative humidities?

If drought-resistant varieties do have higher temperatures than drought-sensitive varieties, the varieties with high temperatures could then be isolated and incorporated into a breeding program aimed at developing drought-resistant crops. Because water conservation is of top priority in Kansas, it is essential to be able to identify varieties that will yield well with limited amounts of water.

Water Quality

Nitrate Studies

The most critical water quality problem in Nebraska is nitrate contamination (26). It is also a major concern in Kansas (17). In some Nebraska counties, particularly those along the Platte River, the concentration of nitrate-nitrogen in wells has risen from less than 2 parts per million in the early 1960s to more than 30 parts per million. The Public Health Service's drinking water standard is 10 parts per million. The increase is attributed to shallow groundwater levels, medium- to coarse-textured soils, shallow soils, high nitrogen fertilizer rates, and excess water from irrigation.[2]

The high nitrate concentrations are a health problem because of the possibility of methemoglobinemia. In this disease, nitrate is converted to nitrite prior to or during the ingestion and absorption of nitrate. Nitrite then changes oxyhemoglobin to methemoglobin, resulting in a lack of oxygen in the blood. Babies are more susceptible to the disease than adults (21). Infants on farms in the North Central Region have suffered methemoglobinemia poisoning because of nitrates in well water (47).

One of the most important factors affecting nitrate concentrations in soils is the presence or absence of plants. Fallow soils often have higher nitrate levels than cropped soils. A crop cover on the land takes up nitrate, reduces runoff, and keeps the soil loose for good percolation. Most crops take up nitrogen roughly in proportion to their yield. Some crops, however, accumulate unusually high concentrations of nitrate. The ability of certain plants, such as sudangrass [Sorghum bicolor, formerly Sorghum sudanese (Piper) Stapf)] and Rhodes grass (Chloris gayana Kunth), to mine nitrates is well known. These plants efficiently remove nitrate from the soil and prevent deep seepage and groundwater pollution (21).

The moisture requirement for maximum uptake of nitrate by the nitrate accumulators is not known. High nitrate contents in plants often are associated with droughty conditions because a lack of moisture interferes with normal plant growth and can result in nitrate accumulation in plants. Nitrates are most likely to accumulate during hot, dry weather, when transpiration from plants is rapid and the supply of soil moisture is inadequate to meet the high water demand. Under these conditions,

[2]Frank, Kenneth D. 1982. "The Hall County (Nebraska) Water Quality Special Project." Seminar, October 6, Department of Agronomy, Kansas State University.

plants wilt and stop growth, and nitrates accumulate. Plant roots, however, cannot absorb nitrates from a dry soil. Studies have shown that the greater the water supply, the more rapid and complete the recovery of nitrogen fertilizer is. But if moisture levels are too high, crop yields and nitrate uptake can be depressed. Thus, some optimum soil moisture level should lead to a maximum uptake of nitrates. To reduce nitrate concentrations in soil and groundwater, therefore, specific questions need to be answered:

1. What is the water requirement for maximum nitrate uptake by nitrate-accumulators?

2. Do certain crop varieties or hybrids commonly grown in the North Central Region, for example, corn, soybeans, and wheat, accumulate relatively high amounts of nitrates under specific irrigation treatments?

Answers to these questions would permit identification of the type of plant and water regime that result in maximum nitrate uptake by plants. Data should help economists develop water production functions (14).

Use of Liquid Sewage Sludge and Wastewater

Much research is being done at many experiment stations on the use of sewage sludge and wastewater on land. For example, a study on a Manhattan, Kansas, farm, on which all of the city's sludge has been injected into the soil since 1976, showed that the sludge was a good source of nutrients and water for the crops (corn, sorghum, and winter wheat) (20). A neighboring farm, which received only inorganic fertilizer, served as the control. Concentrations of elements in the sludge-treated plants, including trace elements (cadmium, chromium, copper, nickel, lead, and zinc), were within normal concentration ranges. Soil water content on the sludge farm was greater than that on the control farm. Grain yields with sludge were the same as those with inorganic fertilizer.

Wastewater (secondary effluent) also is a good source of water and nutrients, and there appear to be few physical, chemical, or biological problems associated with its application on agricultural land (21). These studies and others show that sewage sludge and wastewater can be used effectively to recycle nutrients and water for crop growth.

Use of sewage sludge and wastewater on farms is needed. Traditionally, the U.S. public has been reluctant to use human wastes on land. In Israel, a country with severe water shortages, wastewater is the only remaining source of supplemental water (38). More municipal wastewater is used in agriculture in Israel than in any other country. "Under such severe conditions of scarcity as exist in Israel it is unthinkable to allow wastewater from urban centers to be lost to the water economy by flowing to the sea" (38). In the United States, as water becomes more scarce, perhaps farmers will be more willing to use wastewater and sludge on land than they have been in the past.

Basic Research

Development of Drought-Resistant Plants Using Tissue Culture

There is currently much interest in the use of tissue culture to select improved varieties of plants (25, 46). The Evapotranspiration Laboratory is cooperating with the wheat-breeding group at Kansas State University to investigate the possibility of developing drought-resistant wheat plants using tissue culture. We do know that varieties of wheat vary in ease of regeneration (37). For example, a drought-sensitive variety ('Ponca') grew in tissue culture more easily than a drought-resistant variety ('KanKing') (unpublished data). The percentages of callus formation from roots and shoots for the drought-resistant variety were 35 and 32 percent, respectively, and for the drought-sensitive variety, 51 and 46 percent, respectively.

In particular, the study is looking at the production of drought-resistant wheat plants from callus tissue treated with abscisic acid. Abscisic acid is a plant hormone that increases when plants are exposed to drought. It closes stomata, and plants with higher levels of abscisic acid appear to be better able to withstand drought than plants with lower concentrations. By selection for abscisic-acid resistance *in vitro*, wheat plants perhaps can be regenerated that are more drought-tolerant than those not selected for abscisic-acid resistance. Results to date (unpublished data) show that plants developed from callus exposed to abscisic acid have a different water balance than plants regenerated from callus not treated with abscisic acid (Figure 1).

The research suggests that plant scientists need to answer the following questions:

1. Do varieties known to be drought-sensitive regenerate more readily in tissue culture than varieties known to be drought-resistant? If so, growth of crops from tissue culture might be more successful if drought-sensitive varieties were used rather than drought-resistant varieties.

2. Can abscisic acid be used *in vitro* to select drought-resistant plants?

3. If drought resistance can be obtained *in vitro,* can it be transferred to progeny?

4. If drought-resistant plants can be identified in the laboratory *in vitro,* will the drought resistance still be observed under field conditions?

The answer to the last question is extremely important. I know of no plants, selected for a particular characteristic *in vitro,* that have been successfully propagated in the field and shown to possess the selected character in the field. For example, in California, salt-tolerant alfalfa has been developed *in vitro*. But the regenerated plants were weak (*46*) and salt-tolerant plants developed *in vitro* have not yet been planted in the field. Practical applications from tissue culture research will probably not be immediate.

Mechanism of Drought Tolerance

Plant hormones appear to function at the gene level (*39*). Therefore, to understand the basic mechanism of drought tolerance, research must

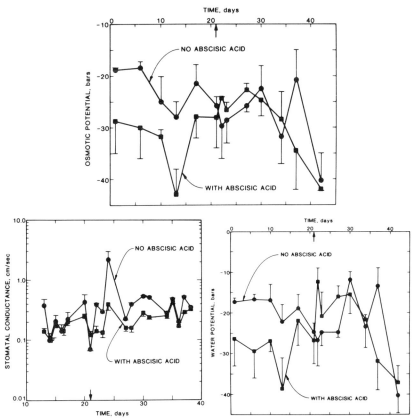

Figure 1. Water relations of winter wheat plants ('ND 7532') regenerated from callus treated with 2 parts per million abscisic acid or with no abscisic acid. Vertical lines indicate standard deviations. Only half the standard-deviation line has been drawn to avoid cluttering the figure. Soil in which plants grew contained equal amounts of sand, peat, and clay loam and was watered every day except day 21 (see arrow) when the soil became dry. The lower water potential, osmotic potential, and stomatal conductance of the plants treated with abscisic acid indicate that the plants might be more drought resistant than the plants not treated with abscisic acid.

focus on plant hormones, particularly the gas ethylene because it is evolved when plants are water-stressed (*1*). Scientists at Kansas State University have found differences between two varieties of winter wheat in ethylene produced by leaves. A drought-resistant variety evolves more ethylene than a drought-sensitive variety (unpublished data). These results are the first to show that drought-resistant and drought-sensitive varieties vary in production of ethylene.

Plant scientists need to pursue the study of this biological phenomenon. Specifically, the following questions need to be answered:

1. Do leaves of drought-resistant plants, in general, produce more ethylene than leaves of drought-sensitive plants?

2. Do species and varieties differ in the amount of ethylene produced?

3. Can drought-resistant varieties be identified using ethylene?

4. What is the mechanism of ethylene in a plant's response to a water deficit? Does it affect stomatal opening, hence, transpiration? Does it affect rate of leaf expansion, hence, the amount of surface area available for loss of water by transpiration? Does it affect rate of root growth, hence, the amount of water available for plant uptake?

If part of the mechanism of drought tolerance is ethylene evolution, then a method will be available for distinguishing drought-sensitive and drought-resistant varieties. Using ethylene to screen for drought tolerance would be relatively easy because analysis of a gas by gas chromatography is routine. Hundreds of varieties could be analyzed in a week. The drought-resistant ones could be identified and planted where water might limit growth. They would survive more readily than the drought-sensitive ones. This is basic research that could be done immediately and should have direct, practical benefits in a short time.

REFERENCES

1. Apelbaum, A., and S. F. Yang. 1981. *Biosynthesis of stress ethylene induced by water deficit.* Plant Physiology 68: 594-596.
2. Black, C. C., Jr. 1973. *Photosynthetic carbon fixation in relation to net CO_2 uptake.* Annual Review of Plant Physiology 24: 253-286.
3. Briggs, L. J., and H. L. Shantz. 1913. *The water requirement of plants. I. Investigations in the Great Plains in 1910 and 1911.* Bulletin No. 284. Bureau of Plant Industry, U.S. Department of Agriculture, Washington, D.C. 49 pp.
4. Briggs, L. J., and H. L. Shantz. 1913. *The water requirement of plants. II. A review of the literature.* Bulletin No. 285. Bureau of Plant Industry, U.S. Department of Agriculture, Washington, D.C. 96 pp.
5. Brown, L. R. 1981. *World population growth, soil erosion, and food security.* Science 214: 995-1002.
6. Campbell, H. W. 1907. *Campbell's 1907 soil culture manual. A complete guide to scientific agriculture as adapted to the semi-arid regions.* H.W. Campbell, Publisher, Lincoln, Nebraska. 320 pp.
7. Cleland, R. 1967. *A dual role of turgor pressure in auxin-induced cell elongation in* Avena *coleoptiles.* Planta 77: 182-191.
8. Cook, R. J., and R. I. Papendick. 1972. *Influence of water potential of soils and plants on root disease.* Annual Review of Phytopathology 10: 349-374.

9. Council for Agricultural Science and Technology. 1982. *Water use in agriculture: Now and for the future.* Report No. 95. Ames, Iowa. 28 pp.
10. Davenport, D. C., and R. M. Hagan. 1981. *Agricultural water conservation in simplified perspective.* California Agriculture 35(11-12): 7-11.
11. Fehrenbacher, J. B., and H. J. Snider. 1954. *Corn root penetration in Muscatine, Elliott and Cisne soils.* Soil Science 77: 281-291.
12. Gardner, W. R. 1973. *Internal water status and plant response in relation to the external water regime.* In R. O. Slatyer [editor] *Plant Response to Climatic Factors.* U.N. Educational, Scientific, and Cultural Organization, Paris, France. pp. 221-225.
13. Hattendorf, M. J. 1982. *Evapotranspiration relationships and crop coefficient curves of irrigated field crops.* M.S. thesis. Kansas State University, Manhattan. 116 pp.
14. Hexem, R. W., and E. O. Heady. 1978. *Water production functions for irrigated agriculture.* Iowa State University Press, Ames. 215 pp.
15. Holbrook, F. S., and J. R. Welsh. 1980. *Soil-water use by semidwarf and tall winter wheat cultivars under dryland field conditions.* Crop Science 20: 244-246.
16. International Soil Tillage Research Organization. 1982. *Proceedings of the Ninth Conference.* Poljoprivredni Institut (Agricultural Institute), Osijek, Yugoslavia. 698 pp.
17. Kansas Water Resources Research Institute. 1980. *Five Year Research Plan, 1982-1987.* Report No. PB81-124505. National Technical Information Service, Springfield, Virginia.
18. Kansas State University Water Task Force. 1978. *Research and extension solutions to water problems.* Manhattan, Kansas. 28 pp.
19. Kirkham, Don. 1970. *The importance of water resources research.* In P. J. Horick [editor] *Water Resources of Iowa.* University Printing Service, Iowa City, Iowa. p. 158-175.
20. Kirkham, M. B. 1982. *Appropriate technology of alternative agricultural practices on drylands.* Final Report. Grant No. NSF/ISP-8014715. National Technical Information Service, Springfield, Virginia.
21. Kirkham, M. B. 1982. *Problems of using wastewater on vegetable crops.* In Proceedings, Workshop on Wastewater Irrigation of Horticultural Crops. American Society of Horticultural Science, Arlington, Virginia.
22. Mac Key, J. 1980. *Shoot:root interrelation in cereals and grasses.* In *Plant Roots. A Compilation of Ten Seminars.* Department of Agronomy, Iowa State University, Ames. p. 29-51.
23. Meyer, B.S., D. B. Anderson, and R. H. Bohning. 1960. *Introduction to plant physiology.* D. van Nostrand Co., Inc., Princeton, New Jersey. 541 pp.
24. Mtui, T. A., E. T. Kanemasu, and C. Wassom. 1981. *Canopy temperatures, water use, and water use efficiency of corn genotypes.* Agronomy Journal 73: 639-643.
25. National Science Foundation. 1982. *The plant sciences. A mosaic special.* Mosaic 13(3): 1-52.
26. Nebraska Water Resources Center. 1981. *Workshop on Nebraska water problems.* Institute of Agriculture and Natural Resources, University of Nebraska, Lincoln. 13 pp.
27. Nebraska Water Resources Center. 1982. *Workshop on water research needs.* Institute of Agriculture and Natural Resources, University of Nebraska, Lincoln. 27 pp.
28. Nelsen, C. E., and G. R. Safir. 1982. *Increased drought tolerance of mycorrhizal onion plants caused by improved phosphorus nutrition.* Planta 154: 407-413.
29. Pepe, J. F., and J. R. Welsh. 1979. *Soil water depletion patterns under dryland field conditions of closely related height lines of winter wheat.* Crop Science 19: 677-680.
30. Phillips, R. E., and Don Kirkham. 1962. *Soil compaction in the field and corn growth.* Agronomy Journal 54: 29-34.
31. Powledge, F. 1982. *Water. The nature, uses, and future of our most precious and abused resource.* Farrar Straus Giroux, New York, New York. 423 pp.
32. Reicosky, D. C., T. C. Kaspar, and H. M. Taylor. 1982. *Diurnal relationship between evapotranspiration and leaf water potential of field-grown soybeans.* Agronomy Journal 74: 677-673.
33. Rose, E., and M. B. Kirkham. 1982. *Root growth of a tall and a short winter wheat.* Kansas Water News (in press).

34. Safir, G. R. 1980. *Vesicular-arbuscular mycorrhizae and crop productivity.* In P. S. Carlson [editor] *The Biology of Crop Productivity.* Academic Press, New York, New York. pp. 231-252.

35. Safir, G. R., and C. E. Nelsen. 1982. *Water and nutrient uptake by vesicular-arbuscular mycorrhizal plants.* In *Role of Mycorrhizal Associations in Crop Production.* Rutgers University, New Brunswick, New Jersey (in press).

36. Salisbury, F. B., and C. W. Ross. 1978. *Plant Physiology.* Wadsworth Publication Co., Inc., Belmont, California. 436 pp.

37. Sears, R. G., and E. L. Deckard. 1982. *Tissue culture variability in wheat: Callus induction and plant regeneration.* Crop Science 22: 546-550.

38. Shuval, H. I. 1982. *The impending water crisis in Israel.* In *Selected papers on the Environment in Israel.* No. 9. Environmental Protection Service, Ministry of Interior, Jerusalem, Israel. p. 43-56.

39. Skoog, F., and R. Y. Schmitz. 1972. *Cytokinins.* In F. C. Steward [editor] *Plant Physiology. Vol. VIB: Physiology of Development: The Hormones.* Academic Press, New York, New York. p. 181-243.

40. Smith, F. W., W. L. Powers, G. E. Smith, R. Koob, J. L. Wiersma, C. Philips, W. A. Hunt, and M. D. Dougal. 1980. *Regional five-year water research and development. Goals and objectives. Missouri Water Resources Research Institutes, Missouri River Basin States.* Office of Water Research and Technology, U.S. Department of the Interior, Washington, D.C.

41. Suksayretrup, K., M. B. Kirkham, and H. C. Young, Jr. 1982. *Stomatal resistance of five cultivars of winter wheat infected with leaf rust (*Puccinia recondita f. sp. *tritici).* Journal of Agronomy and Crop Science 151: 118-127.

42. Tanner, C. B., and T. R. Sinclair. 1982. *Efficient water use in crop production: Research or re-search.* In H. M. Taylor and W. R. Jordan [editors] *Efficient Water Use in Crop Production.* American Society of Agronomy, Madison, Wisconsin.

43. United States Department of Agriculture and the State Universities and Land Grant Colleges. 1969. *A national program of research for water and watersheds.* Research Program Development and Evaluation Staff, Washington, D.C. 99 pp.

44. van den Honert, T. H. 1948. *Water transport in plants as a catenary process.* Discussions Faraday Society 3: 146-153.

45. van Schilfgaarde, J., and G. J. Kritz. 1981. *Water. A basic resource.* In W. E. Larson, L. M. Walsh, B. A. Stewart, and D. H. Boelter [editors] *Soil and Water Resources: Research Priorities for the Nation.* Soil Science Society of America, Madison, Wisconsin. p. 1-19.

46. Venne, R. V., editor. 1982. *Special issue: Genetic engineering of plants.* California Agriculture 36(8): 1-36.

47. Washington, D.C. Evening Star 1968. *Scientist warns of nitrogen peril in food, water.* Friday, December 27, 1968.

48. Weaver, J. E. 1968. *Prairie plants and their environment. A fifty-year study in the Midwest.* University of Nebraska Press, Lincoln. 276 pp.

49. Widtsoe, J. A. 1911. *Dry-farming. A system of agriculture for countries under a low rainfall.* The Macmillan Co., New York, New York. 445 pp.

50. Wittwer, S. 1982. *New technology, agricultural productivity, and conservation.* In H. G. Halcrow, E. O. Heady, and M. L. Cotner [editors] *Soil Conservation Policies, Institutions, and Incentives.* Soil Conservation Society of America, Ankeny, Iowa. pp. 201-215.

10

Water Resources Research: Potential Contributions by Agricultural Engineers and Hydrologists

E. C. Stegman, J. R. Gilley, and M. E. Jensen

As global population increases, food and fiber demands will put added pressure on soil and water resources. Estimates indicate that by 2030 the United States will need about an 85 percent increase in food and fiber production relative to 1977 levels (*25*). Such an increase is possible if past growth rates in agricultural productivity continue. Increases in productivity have varied from 2.1 percent annually between 1939 and 1965 to about 1.7 percent in recent years (*3*).

Growth rates, however, may be plateauing for some crops, and many analysts contend that productivity improvements have resulted from technologies that are too exploitive of natural resource inputs (*24*). Erosion rates are reportedly lowering inherent soil productivity on about one-fourth of U.S. cropland (*25*). Groundwater supplies are being mined. Intensive agriculture also produces other side effects, including pesticide residues and water pollution.

The climate in the North Central Region ranges from semiarid on its western edge to humid in the eastern section. The region encompasses all or parts of the Missouri, Souris-Red-Rainy, Upper Mississippi, Great Lakes, and Ohio water resource regions. The U.S. Water Resources Council (*28*) has identified the following water resource problems in the region: (1) inadequate surface water supply, (2) overdraft of groundwater, (3) pollution of surface and groundwater, (4) quality of drinking water, (5) flooding, (6) erosion and sedimentation, (7) dredging and disposal of dredged material, (8) wet soils and wetlands, and (9) degradation of bay, estuary, and coastal waters. In the following discussion of research needs, we have chosen headings from this list that have particular relevance to agriculture in the North Central Region.

Inadequate Surface Water Supply

Water supply, as characterized by annual surface runoff, averages less than 1 inch over much of North Dakota, South Dakota, and Nebraska to more than 7 inches in southern Illinois, Indiana, and Ohio (28). Evaporative potentials in the semiarid portion of the Missouri resource region greatly exceed growing season precipitation, such that water deficits limit plant growth and yield almost every year. Also, despite lesser evaporation-precipitation differentials in the humid portions of the North Central Region, short period droughts occur frequently.

The technical resolutions for inadequate surface water supply generally fall into two categories: methods that increase available supply and methods that reduce demand.

Increasing Water Supply

Irrigation is agriculture's most significant attempt to augment water supplies. Irrigated area has increased substantially in most states of the North Central Region. For example, from 1975 to 1979, increases ranged from 33 percent in Nebraska to 202 percent in Michigan (14). For the entire region the increase in this period averaged 44 percent on an acreage basis. Sprinkler systems serve almost all of the newer irrigated area: 64 percent are center pivot designs and 16 percent are "big-gun" or travelling linear units.

In the irrigated areas, surface water supplies are, in effect, being increased by pumping from groundwater aquifers. Wells supply about 63, 77, and 90 percent of the irrigation systems in the Great Lakes, Upper Mississippi and Ohio, and Missouri resource regions, respectively. With this recent surge in groundwater use for irrigation, there is a pressing need for improved aquifer data. Models need to be developed and calibrated for major aquifers and then used as tools to predict likely consequences of development scenarios. Where possible, these models should be sufficiently complete to allow analyses of groundwater and surface water interactions (i.e., conjunctive use potential) and in-stream flow requirements.

Major surface water impoundments are located on the Missouri River in North and South Dakota. Diversions to large areas of irrigable soils are an intended use for a significant portion of the water, according to the Pick-Sloan plan. The Oahe project in South Dakota is, however, in a deauthorized state. Distribution works for the Garrison unit in North Dakota are about 20 percent completed.

Numerous court injunctions favoring environmental factions have plagued recent progress. Environmental concerns have run the gamut, from those of inadequate benefit/cost ratios to the potential for disastrous transfers of foreign biota from Missouri waters, via project return flows, to the Souris-Red-Rainy drainage basin. Future progress of this project may depend in part on whether "best management practices" prove effective for limiting nutrient and dissolved solids losses to acceptable levels.

An alternative frequently considered for increasing water supplies is to transfer water from areas with a surplus to areas with deficits. Water transfers have been practiced in the western United States for nearly a century (11), but future transfers will be more difficult for two reasons. First, political and environmental aspects will create lengthy delays. Second, the costs of future water transfer projects will be extremely high.

Subregional water importation systems and large-scale interbasin water transfer systems are among the alternative choices being considered by the High Plains study council for partially offsetting future impacts of groundwater declines in the Ogallala Aquifer. The major physical obstacle is high pumping lift. Much of the High Plains is above 4,000 feet mean sea level; thus, lifts from the Missouri and Mississippi drainage basins would exceed 2,500 to 3,000 feet. The energy required for this lift appears to exceed the economic capacity of agriculture to pay for the water. But, given the fact that these systems are being considered, there is the need for each state to better define or predict its future water needs. Interestingly, from an agricultural perspective, some theoretical (22) and applied analyses (21) suggest that crop transpiration efficiencies increase as climatic evaporative demand drops. Thus, higher crop water use efficiency might be obtained by putting greater emphasis on water management and irrigation technology in the Northern and Central Plains, where surface water supplies would also require lower developmental costs.

Reducing Water Demands

In dryland agriculture the challenge is to develop conservation systems that make efficient use of in-situ water supplies. Practices having use in the North Central Region include conservation tillage, snow management, land shaping, water table control (given a low salinity hazard), weed control with herbicides, and runoff storage in on-farm reservoirs.

These practices have potential impacts on water use efficiency (yield per unit water consumed).[1]. The research challenge appears to be one of

[1] Jensen, M. E. 1982. "Agricultural Technology for Water Conservation." Presented at American Association for Advancement of Science Symposium, Water-Related Technologies for Sustaining Agriculture in U.S. Arid and Semi-Arid Lands, Washington, D.C.

developing innovative combinations of these practices that further improve water use efficiencies and that have significant economic appeal. Indirect technologies may also increase water use efficiency, such as plant breeding, fertilizer application, and plant population.

Overdraft of Groundwater

Groundwater resources, as indicated earlier, represent the primary water supply for irrigation in the North Central Region. When used for this purpose, it is estimated generally that evapotranspiration uses nearly 80 percent of the pumped water (15). Significant groundwater mining is taking place in the Southern Great Plains and to lesser extent in the Upper Republican, Upper Big Blue and Upper Little Blue River Basins in Nebraska (10). Given the low natural recharge potential in most of the Northern Great Plains, it appears likely that the groundwater overdrafts will spread in areal extent as irrigation development increases. Groundwater aquifers in humid parts of the North Central Region have a much higher natural recharge potential, but greatly increased demands for irrigation could eventually cause some overdraft problems in these areas as well (29).

In overdraft areas the alternatives, assuming little or no water importation, are essentially (1) let groundwater development and eventual decline run its course, (2) curtail future development, (3) control the present uses of groundwater, and (4) develop and implement cost-effective conservation practices. These four courses of action are already taking place in varying degrees. Lands are reverting to dryland in the Southern Great Plains, largely for economic reasons. Water permit systems in some states allow what are estimated to be sustainable levels of development. When this level is reached, no further permits are granted until monitoring programs indicate that further development is sustainable. Legislation in Nebraska has established groundwater control areas in which associated Natural Resources Districts are authorized to adopt groundwater management regulations. Experience with regulatory measures has, however, generally been too brief for effects to be very evident (10).

Alternative 4 translates into the complex matter of attaining acceptable returns with limited use of irrigation water. Due to rising energy costs, farmer acceptance of the concept of limited irrigation is increasing. However, this squeeze on profit margins frequently necessitates improvements in productivity or production efficiency also.

Many of the research needs for limited irrigation practices fall within the domain of irrigation scheduling, which implies the use of techniques for determining when to irrigate and the amount of water to apply during each period of irrigation.

Irrigation Scheduling

Emphasis in the past 10 years on irrigation scheduling has been directed mainly at applied water balance methods. A widely used procedure is a U.S. Department of Agriculture computer program developed by Jensen and associates (*16, 17*). This program, with local modifications, is used largely by consulting firms, which, in addition to irrigation scheduling, often provide services in the areas of plant nutrition, pest management, agronomic guidance, and engineering to improve system operations.

Water balance reliability depends upon the accuracy with which the components (crop evapotranspiration, rainfall effectiveness, irrigation efficiency, deep percolation, and plant available soil water storage capacity) can be estimated. Present water balance procedures are not sufficiently accurate to obviate the need for extensive field monitoring of soil water to check water balance predictions. Thus, improved techniques are needed. Principal problems with present scheduling models include:

1. Procedures for estimating evapotranspiration make no allowances for plant population, crop residue, or row spacing effects.

2. Phenologic models as functions of climatic variables and stress management are inadequately defined.

3. Procedures are inadequate for estimating the degree of plant water stress and its effect upon water use.

4. Rooting models are inadequately defined.

5. Methods for estimating the effectiveness of precipitation and deep percolation are inadequate.

6. Precipitation forecasts and probabilities are inadequately used.

On the assumption that water balance models can be developed to have an adequate capability for estimating prevailing soil water deficits, there is still the need to improve methods for deciding when to irrigate. Many studies have produced semiquantitative criteria for timing of irrigation amounts. Usually, these criteria are in the form of parameter thresholds, such as allowable soil water depletion, allowable evapotranspiration deficit, threshold leaf water potential, and foliage-air temperature difference (*20*). Allowable soil water depletions are the criteria used most often, but to minimize the potential for water stress during high evaporative demand periods these criteria tend to start irrigations at conservative levels of depletion.

Refinements in timing criteria will require elaboration of improved quantitative methods for estimating degree of water stress and for assessing associated consequences on the current crop growth rate and the eventual economic yield. This capability will require further development of applied crop growth models that can be run on computers having access to real-time weather data. Given this information, the first seasonal irrigation could be delayed until water stress approaches a level that affects yield potential by an adverse economic amount. The irrigation season would be terminated when marginal costs of further water application exceed the value of further yield improvement.

Irrigation scheduling concepts have tended to focus only on growing season management of irrigation water applications. In much of the North Central Region a more integrated approach to irrigation is needed that focuses on year-round water management concepts, combining conservation tillage, residue management, irrigation scheduling, and other ideas to achieve technologies that minimize annual irrigation amounts by reducing soil-surface evaporation, deep percolation, and runoff.

On-Farm Computer Applications

Emerging computer technology, we believe, will lead to many on-farm uses that will directly or indirectly bring about improvements in water conservation or water use efficiency. It is predicted that computers will soon become an indispensible "hired hand" on America's farms (23). Farmers are already using microcomputers for accounting, acquiring market information, and making management decisions with available software on systems, such as AGNET. With the virtual explosion in new hardware currently taking place in the microprocessor and microcomputer industry, a vast proliferation in new applications can be expected. Not only will there be much software generated to aid decision-making, but there will also be new opportunities for process monitoring and control.

In irrigation scheduling, microcomputer-based systems will likely perform the functions of data acquisition, simulation model (i.e., water balance and crop growth) calculations, and direct irrigation system control. Furthermore, sensors in the field or on the system will feed back information to the computer for real-time analysis and on-the-spot adjustment of system operation.

It is envisioned that in the next decade whole farm management models will be available that (1) manage all farming operations, (2) allocate labor and resources to each enterprise, (3) monitor enterprise operations, (4) update data bases that are used in recordkeeping and management operations, and (5) project potential impacts of alternative courses of enterprise management (23). Modeling needs for the development of comprehensive management systems cut across almost all disciplines normally associated with agriculture.

Water Quality

The primary nonpoint source water pollutants from agriculture are sediment, nutrients, and pesticides.

Erosion and Sedimentation

Erosion and sedimentation continue to be widespread problems in the North Central Region. Severity is greatest in humid areas, where soil losses are estimated to exceed 5 tons per acre on 27 percent of tilled cropland (*29*). Erosion problems also tend to be inadequately recognized because technologies for productivity improvement have masked or offset erosion impacts (*29*).

Surface cover is the most important factor in reducing soil losses on agricultural lands (*30*). Most studies show that conservation tillage reduces erosion from 50 to 90 percent when compared with conventional tillage (*6*). The cropland area under conservation tillage practices has increased substantially in recent years as farmers have sought to reduce production costs.

Soil drainage is a key factor affecting the economic feasibility of conservation tillage in the Corn Belt states of Ohio, Indiana, Illinois, and Iowa (*6*).

The semiarid climate of the Great Plains favors conservation tillage for its moisture-conserving potential. The percentage of land in conservation tillage declines from south to north, primarily because residue cover lowers soil temperatures in the spring. Conservation tillage makes double cropping possible in the southern part of the North Central Region, thus providing considerable economic incentive for adoption. Innovations that achieve earlier planting and harvest of the first crop in a double crop sequence could expand this practice further north.

To increase the rate of adoption of conservation tillage and the percentage of land in conservation tillage, systems need to be developed that are attuned to each type of farming environment. Deterrents such as problems with soil drainage, seed zone soil temperatures, weed control, disease control, and seed and fertilizer placement can likely be reduced through interdisciplinary approaches, including new equipment designs.

Attempts to quantify water erosion processes range from the well-known universal soil loss equation (USLE) to more recent "distributed parameter" hydrologic models (*11*). The USLE, although widely accepted and used, is an empirical procedure that is useful primarily for estimating average annual soil loss rates from field-size watersheds. This method includes a cover-management factor to account for conservation tillage or residue management effects. Cover management factors (C-values) are available, but further definition and refinement are needed for recent developments in conservation tillage methods.[2]

Water quality control planning, prompted by the Federal Water Pollu-

---

[2]Personal communication with E. C. Dickey, Agricultural Engineering Department, University of Nebraska, Lincoln.

tion Control Act Amendments (Public Law 92-500), requires a different look at erosion and sedimentation problems. Among the law's requirements are (a) an assessment of the source and amount of sediment by type of land use, (b) economic evaluation of the costs and benefits of erosion and sediment control procedures, and (c) evaluation of the effects of a mix of land uses on water quality.

To address the requirements of the 208 planning section of Public Law 92-500, much recent research has focused on the development of more complete erosion simulation models. One of the most useful tools in the complex planning of nonpoint source pollution control programs is an accurate, comprehensive watershed model for simulating the effects of hypothetical control measures (1).

Distributed parameter models attempt to preserve and use information concerning the areal distribution of spatially nonuniform processes. These models also allow planners to be site specific in developing and evaluating the effectiveness of potential control measures. No single model currently available can handle the entire range of complicated nonpoint source pollution problems, but several are approaching a state of development that can make them effective planning tools (1). Further research is needed to (1) improve mathematical models capable of simulating erosion and hydrologic processes, (2) quantify the parameter values required in these models to simulate the behavior of a broad spectrum of naturally occurring conditions, and (3) quantify the effects of various best management practices for improving water quality from agricultural runoff and drainage on small and large areas (12).

Comprehensive erosion transport and deposition models of the distributed-parameter type have voracious data requirements. A large store of information exists on soil properties and performance characteristics for thousands of soils, but it is scattered in numerous publications and reports. To calibrate and use hydrologic models, there is a particularly acute need for the development of computerized data banks. Such banks should contain all information on the physical, chemical, and biological properties of named soils and be easily accessible (9).

Nutrients

The principal concerns with nutrients from agriculture in the environment are the effects of nitrogen in water on human health and of nitrates and phosphorous in accelerating eutrophication of water bodies. Excessive precipitation in the Corn Belt causes significant surface runoff and flow to subsurface drainage systems. Many studies (28) of nutrient loss confirm that agriculture contributes significant amounts of nitrogen and phosphorus to surface waters. In some irrigated areas of Nebraska, there is evidence of substantial nitrogen loss to groundwater (18). For example, it is estimated that nitrate concentrations in groundwater are increasing by 1.1 parts per million per year in Holt County, Nebraska (8). Expanded irrigation development in humid and subhumid areas of the region on coarse-textured soils may represent a serious pollution threat to groundwater, given the unpredictability of rainfall. Current methods are inadequate for using precipitation forecasts or probabilities in the irrigation scheduling process.

Present practices for nutrient input to conventional dryland and irrigated agriculture are based heavily on soil testing criteria. These criteria are in turn the result of local fertilizer-yield testing programs. Although estimated to be reasonably accurate, these criteria nevertheless represent averaged nitrogen-cycle data. Improvements in applied methods await the development of models that better predict the transformations of nitrogen in the whole soil-water-plant-atmosphere environment. Important processes needing further elucidation are those of organic matter mineralization and nitrate denitrification.

Criteria for timing of fertilizer applications are also approximate, particularily for split applications of nitrogen through irrigation systems. Better understanding of crop uptake cycles and rates of nutrient release from soils and applied fertilizers are needed.

Fertilizer application methods are in need of engineering improvements. Current practices typically result in a given rate of fertilizer application to field-size areas with little regard for soil spatial variability. Innovations in computer technology could lead to microprocessor-microcomputer systems that translate inputs of soil-mapping data to the control of equipment in fields, such that fertilizer application amounts are tailored to differing land parcels or soil types. Similar innovations are likely for irrigation systems. For example, fertilizer, herbicide, and water application (common or separate units) apparatus could be developed for center pivot and linear irrigation systems. Under the direction of on-board computer systems it would then be possible to tailor applications of fertilizers, pesticides, and water amounts to spatial variabilities of soils, weeds, diseases, etc. High travel speed and low application rates would further allow the flexibilities of operation that are needed in humid areas.

Current low-pressure application devices for center pivot irrigation systems aggravate water application intensities. Thus, the likelihood for surface runoff, nonuniform application, and sediment or nutrient transport increases. Much remains to be learned about cultural practice, system management, and sprinkler device interactions with regard to problem soils and topographies in specific climates.

Management criteria for nutrient inputs to conservation tillage systems need further evaluation in climatic settings where enhanced water infiltration also creates potential for increased nitrate losses (6).

Nutrient losses are reduced but not eliminated by improved management of nutrient inputs. Thus, hazard abatement methods need to be considered. For example, to what extent can we close agricultural systems to achieve greater degrees of nutrient cycling? Can we develop economically effective nitrate recovery methods for treating drainage waters from agricultural lands?

Salinity

Salinity problems in the North Central Region occur primarily in the Missouri water resource region, particularly in the Dakotas. In some areas, brackish groundwater aquifers are the only dependable source of water supply for domestic, community, and industrial use (*27*). Salinity in return flows from project-scale irrigation development in the Dakotas may increase treatment costs for some municipal facilities (*13*). Other problems include saline seeps in nonirrigated areas and increasing use of groundwater of marginal quality for irrigation. Management and drainage criteria for these waters need refinement, particularly for glacial till soils. For return flows that impact Canada, there may be need to consider re-use on salt tolerant crops, thereby reducing drainage volume, after which the water would be disposed of in salt sinks (*13*).

Pesticides

Pesticides are an integral component of modern crop production practices. Their use in U.S. agriculture has risen rapidly in the past 20 years, increasing by about 40 percent (*7*). Substantial increases are generally projected for the next two decades also, particularly if no-till planting systems, which require heavy use of herbicides, are widely adopted (*24*).

The environmental problems associated with pesticide use include (1) biological amplification and concentration in tissues of higher order predators, including man; (2) development of increased resistance to pesticides by insect pests; (3) destruction of natural pest controls, such as insect-eating birds and predatory insects; (4) emergence of new pests previously not troublesome; and (5) the potential for accidental poisonings (*4*).

A primary program for reducing negative environmental impacts of pesticides has been the U.S. Environmental Protection Agency's development and release of new pesticides. Significant progress has been made in replacing persistent pesticides (organochlorines) with more readily degradable, short-toxic-life compounds. There is, however, an ongoing concern that even low residue levels of some pesticides or their metabolites could act as carcinogens, mutagens, or teratogens.

Methods that increase efficiency of pesticide activity and reduce losses not only reduce potential environmental concerns but also improve the economics of crop production. The primary loss routes of pesticides from treated land include (1) spray drift and volatilization, (2) biological or chemical degradation, (3) plant uptake and harvest removal, (4) adsorption to soil surfaces with removal by erosion, and (5) dissolution in rainwater or irrigation and removal by surface runoff or percolation through the soil to groundwater.

Losses to groundwater are a specific concern, because once groundwater quality is adversely affected, it is difficult to reverse the effect. Research is needed to further elucidate important factors affecting the transport of pollutants from recharge areas to aquifers.[3] Mathematical models need further development, but physical measurements basic to the modeling efforts are also needed. In areas where groundwater is already adversely affected, research is needed on economic methods of removing contaminants, such as pesticides, nitrates, halomethanes, sodium, and radionuclides from drinking water.

Engineering opportunities exist in the development of improved methods for reducing pesticide losses into undesirable pathways. Because most pesticides have a short toxic life and a strong affinity for soil, the greatest potential for undesirable effects occurs in the period immediately following application. Thus, practices that help retain soil and water reduce pesticide losses substantially. Some specific areas requiring further development are (1) conservation tillage and pesticide interactions; (2) models of pest ecologies that result in criteria for optimum timing and dosage and that interface with other management models, such as irrigation scheduling; (3) development of application equipment methods that improve pesticide placement or efficiency resulting in lower rates or reduced frequency of application; (4) methods that reduce drift and optimize deposition onto plant surfaces; and (5) microprocessor systems that implement disease or insect management models with real-time input of weather and associated data from in-field sensors.

Drainage of Wet Soils and Wetlands

Wet Soils

There are times in each growing season in much of the humid part of the North Central Region when agricultural soils are too wet for field operations. Excessive surface and subsurface soil wetness hinders timely operations, such as tillage, planting, chemical applications, and harvesting. Excessive moisture can also reduce crop yields. Farmers overcome

---

[3]Personal communication with J. L. Baker, Agricultural Engineering Department, Iowa State University, Ames.

soil wetness problems by installing surface and subsurface drainage systems. It is estimated that more than 100 million acres of land is drained in the humid regions of the United States (29).

Weather patterns in humid areas of the North Central Region also frequently exhibit 1 to 6 week periods of drought when plant growth is reduced for lack of water. Thus, there is increasing interest in the development of technologies for installation and management of reversible subsurface drainage systems.[4] Shallow, nonsaline water tables are a valuable resource, and there is increasing evidence that drainage designs should preserve this resource for crop use (2). Conserving and managing shallow water tables can also save significant amounts of irrigation water and, consequently, energy that would otherwise be required to obtain maximum crop yields. Research is needed to take a new look at subirrigation as a water management practice. Specifically, there is the need to determine natural conditions, associated system designs, and management procedures that make subirrigation economically attractive.

Recent technological changes have had considerable effect on subsurface drainage. Corrugated plastic drain tubing has changed the traditionally accepted concepts of quality with regard to drains, drain filters, and drain installation. Laser-controlled systems and high-speed trenchers and plows have increased the need for new quality control criteria for plastic tubing and associated filters.[5]

Soil compaction problems appear to be increasing in humid-area soils (26).[6] There are a number of possible causes, including heavier tractors, continuous row crop production, untimely field operations, and increased use of conservation tillage. Research is needed to establish measurable parameters that describe compaction. These parameters can either be included in present water management models or in new models designed to account for the compaction problem in drainage system design.

The greater amounts of surface residue inherent in conservation tillage also reduce soil surface evaporation. During periods of low rainfall, this is desirable. However, during wet spring seasons, the effect allows fewer days for tillage and planting operations. Research is needed to determine if subsurface drainage systems can be designed to improve the effectiveness of minimum tillage by lowering the water content of the trash-covered soil surface when such a reduction is desirable.

One of the most difficult tasks of drainage engineers has always been to evaluate the benefits received from specific drainage installations. Recent advances in remote sensing permit improved yield estimates on a

---

[4]Personal communication with W. Lembke, Agricultural Engineering Department, University of Illinois, Urbana.
[5]Ibid.
[6]Ibid.

field-by-field basis. Research is needed to determine if such yield estimates, taken in successive years prior to, during, and after a drainage system installation, can be used to evaluate the benefits ascribed to drainage.

Significant advances have been made recently in the development of simulation models for evaluating drainage needs and benefits on a probability basis (*19*). A major weakness of such simulations is the general lack of crop response functions. Laboratory, field, and theoretical work are needed to better define dynamic crop growth and economic yield relationships to parameters that describe soil drainage or wetness on a continuous basis.

Most drainage models are of the lumped parameter type, i.e., they use aggregated representations of physical parameters. Further work is needed with distributed parameter concepts to develop techniques that simulate field drainage-crop growth interactions in both a spatial and temporal sense. Given this capability and the use of feedback instrumentation, computer systems could be designed to control the operation of drainage structures or systems.

Wetlands

The issue of wetlands' value versus use of wetlands for agricultural purposes has been a controversial one for many years. Despite recent federal programs designed to preserve and protect remaining wetlands, there is still a need to develop better understanding of their wildlife and hydrologic functions. The contention is that wetlands play important roles in flood damage reduction, erosion control, sediment entrapment, groundwater recharge, and water supply (*27*). Support for this hypothesis is generally site specific. Specific needs in semiarid, water-short areas are for methods that allow quantitative documentation of various wildlife habitat practices on water consumption and cost of reducing their water consumption relative to an increase in wildlife productivity.[7] These methods would permit comparisons of wildlife benefits per unit of water consumed in a manner analogous to terms like crop water use efficiency.

[7]Jensen, M. E., "Agricultural Technology."

REFERENCES

1. Beasley, D. B., L. F. Huggins, and E. J. Monke. 1982. *Modeling sediment yields from agricultural watersheds*. Journal of Soil and Water Conservation 35(3): 113-117.
2. Benz, L. C., E. J. Doering, and G. A. Reichman. 1981. *Watertable management saves water and energy*. Transactions, American Society of Agricultural Engineers 24(4): 995-1,001.
3. Bertrand, A. R. 1981. *Soil and water resources: Research priorities for the future*. In William E. Larson, Leo M. Walsh, B. A. Stewart, and Don H. Boelter [editors] *Soil*

and Water Resources: Research Priorities for the Nation. Soil Science Society of America, Inc., Madison, Wisconsin. pp. 198-207.

4. Caro, J. H. 1976. Pesticides in agricultural runoff. In Control of Water Pollution from Cropland, Volume II, An Overview. ACS-H-5-2 and EPA-600/2-75-0265. U.S. Government Printing Office, Washington, D.C. pp. 91-119.

5. Cosper, H. R. 1979. Soil taxonomy as a guide to economic feasibility of soil tillage systems in reducing nonpoint pollution. Economics, Statistics and Cooperative Service, U.S. Department of Agriculture, Washington, D.C.

6. Crosson, P. 1981. Conservation tillage and conventional tillage: A comparative assessment. Soil Conservation Society of America, Ankeny, Iowa. 35 pp.

7. Eichers, T. R., P. A. Andrilenos, and T. W. Anderson. 1978. Farmers' use of pesticides in 1976. Agricultural Economics Report No. 418. U.S. Department of Agriculture, Washington, D.C. 58 pp.

8. Exner, M. E., and R. F. Spalding. 1979. Evolution of contaminated groundwater in Holt County, Nebraska. Water Resources Research 15(1): 139-147.

9. Flach, K. W., and C. Johnson. 1981. Land resource base and inventory. In William E. Larson, Leo M. Walsh, B. A. Stewart, and Dan H. Boelter [editors] Soil and Water Resources: Research Priorities for the Nation. Soil Science Society of America, Inc., Madison, Wisconsin. pp. 21-39.

10. Gessaman, P. H. 1981. Initial perspectives on Nebraska's state water planning and review process. Report No. 122. Department of Agricultural Economics, University of Nebraska, Lincoln. 30 pp.

11. Haan, C. T., H. P. Johnson, and P. L. Brakensick, editors. 1982. Hydrologic modeling of small watersheds. American Society of Agricultural Engineers, St. Joseph, Michigan.

12. Huggins, L. F., D. B. Beasley, G. R. Foster, S. J. Mahler, and E. J. Monke. 1981. Characterization of the hydrology of small watersheds. Agricultural Engineering Department Annual Report, Purdue University, West Lafayette, Indiana. pp. 22-23.

13. International Joint Commission. 1977. Transboundary implications of the Garrison Diversion Unit. Washington, D.C.

14. Irrigation Journal. 1979. 1979 irrigation survey. 29: 20-27.

15. Jensen, M. E. 1982. Overview—irrigation in U.S. arid and semiarid lands. In OTA assessment, Water-Related Technologies for Sustaining Agriculture in U.S. Arid and Semiarid Lands. Office of Technology Assessment, U.S. Congress, Washington, D.C. 73 pp.

16. Jensen, M. E., D.C.N. Robb, and G. E. Franzoy. 1970. Scheduling irrigations using climatic-crop-soil data. American Society of Civil Engineers, Journal of Irrigation and Drainage Division 96(IR1): 25-38.

17. Jensen, M. E., J. L. Wright, and B. J. Pratt. 1971. Estimating soil moisture depletion from climate, crop, and soil data. Transactions, American Society of Agricultural Engineers 14(5): 954-959.

18. Muir, J., J. S. Boyce, E. C. Seim, P. N. Mosher, E. J. Diebart, and R. A. Olson. 1976. Influence of crop management practices on nutrient movement below the root zone in Nebraska soils. Journal of Environmental Quality 5(3): 255-258.

19. Skaggs, R. W. 1980. A water management model for artificially drained soils. Technical Bulletin No. 267. Biologic and Agricultural Engineering Department, North Carolina State University, Raleigh. 54 pp.

20. Stegman, E. C. 1982. Irrigation scheduling: Applied timing criteria. In D. Hillel [editor] Advances in Irrigation. Academic Press, New York, New York (in press).

21. Stewart, J. I., and R. M. Hagan. 1969. Predicting effects of water shortage on crop yield. Society of Civil Engineers, Journal of Irrigation and Drainage, Division 95(IR1): 94-104.

22. Tanner, C. B., and T. R. Sinclair. 1982. Efficient water use in crop production: Research or re-search. In H. M. Taylor and W. Jordan [editors] Efficient Water Use in Crop Production. American Society of Agronomy, Madison, Wisconsin (in press).

23. Thompson, T. L. 1980. On-farm computer use in 1990. In Proceedings, Program-

mable Calculators and Minicomputers in Agriculture Workshop. Hot Springs, Arkansas. 5 pp.

24. U.S. Council on Environmental Quality. 1980. *The global 2000 report to the president. Technical report, volume two.* U.S. Government Printing Office, Washington, D.C.

25. U.S. Department of Agriculture. 1980. *Appraisal 1980, Soil and Water Resources Conservation Act, review draft.* Washington, D.C.

26. U.S. Department of Agriculture. 1981. *Report of ARS national workshop on soil compaction.* Agricultural Research Service, USDA, Washington, D.C.

27. U.S. Department of the Interior, Office of Water Research and Technology. 1980. *Regional five-year water research and development goals and objectives, Missouri Water Resources Research Institutes.* Washington, D.C.

28. U.S. Department of the Interior, Water Resources Council. 1978. *The nation's water resources. The second national water assessment. Volume 2: Water quantity, quality and related considerations.* Washington, D.C. 134 pp.

29. VanDoren, D. M. 1981. *Conservation needs and technological on agriculture land.* In William E. Larson, Leo M. Walsh, B. A. Stewart, and Dan H. Boelter [editors] *Soil and Water Resources: Research Priorities for the Nation.* Soil Science Society of America, Inc., Madison, Wisconsin. pp. 41-61.

30. Wischmeier, W. H. 1973. *Conservation tillage to control water erosion.* In *Conservation Tillage.* Soil Conservation Society of America, Ankeny, Iowa. pp. 5-12.

11

Water Resources Research: Potential Contributions by Economists

John F. Timmons

I begin with eight considerations relating to water:

First is the recognition of the heterogeneity of water supply and water demand conditions throughout the North Central Region. The major unifying water characteristic in the region consists of the common rivers and aquifers carrying supplies of water to, from, and through the region. But these water supplies also involve and affect states or parts of states in other regions.

Second, water joins food and air as the three essential sustenance resources for human existence as well as for the growth of plants and animals that provide us with food and fiber. Beyond mere existence, water provides us with recreation, energy, transportation, sewage disposal, and many other goods and services necessary for the pursuit of happiness.

Third, in providing these goods and services, water serves as an essential resource for satisfying both derived and direct demands of people.

Fourth, water is put to consumptive uses as well as nonconsumptive uses in providing goods and services for people.

Fifth, the nature of water is complicated by its detrimental effects as well as beneficial effects upon people and the natural environment.

Sixth, supplies of water are confounded by both exhaustible and non-renewable as well as flow and renewable sources (*13*).

Seventh, water resources are intimately intertwined with soil resources in crop production as well as with capital investments in the production process, without as well as within the agricultural sector.

Eighth, these supply and demand characteristics bring water into the purview of economics, which is basically concerned with the allocation of scarce resources among competing uses as well as the income distribution effects of such allocations. On the other hand, economics is also concerned with the detrimental affects of water that reduce crop yields, interfere with many human activities, and destroy or damage public and private improvements on and in land.

The minimax concept, as a core of economic theory and practice, applies to the development and allocation of water resources in instances of both scarcity and excessiveness.

Because allocations of water among competing uses are seldom or imperfectly provided by markets, the process of allocation becomes more difficult in achieving an optimum allocation of water among competing uses.

Identification of Continuing and Emerging Problems

In the process of identifying and assessing the potential contribution of economics and economists to research on water problems, it is necessary to identify the nature of continuing and emerging water problems to which economic theory and methods can be applied. This process involves a review of the work by economists. It also includes research that the science of economics might suggest from the theory and methods comprising economics as a discipline, but which may or may not appear in the accomplishments of economists thus far.

In defining water problems, experienced or expected, some relevant optimum is required to delimit gaps between desired and experienced or expected consequences. If no gaps are discovered between the optimum and the consequences, then no problem is delimited in a manner that is amenable to inquiry, at least in terms of the manner in which the optimum and consequences or expectations have been derived and stated. According to Charles Howe (8), "The characterization of a socially optimum pattern of resource use, while of necessity quite abstract, provides a benchmark against which the performance of different market structures can be judged."

Economics provides guides and criteria for identifying this optimum and for evaluating alternative procedures for allocating water and legal and technological means for these allocations, assuming that society is interested in maximizing water's net value product. Problems confronting the realization of this optimum in water allocation and use arise when gaps develop between the optimum and the experienced or expected results.

My review of what economists have been doing in water research, as well as what the discipline of economics implies that economists could be doing, provides nine major categories of water problems inviting contributions by economists. With some overlapping and probably some overlooked or underemphasized problems, these nine categories of developing and emerging problems in the use and management of water resources are as follows: (1) detrimental effects of excess water, (2) increasing scarcities of water, (3) deteriorating water quality, (4) detrimental soil and water interrelationships, (5) externalties, (6) lack of water supply development, (7) exhaustible and nonrenewable characteristics of water sources, (8) allocation of available supplies of water among competing uses and users, and (9) conflicts among states and countries over water ownership, transfer, and use.

These water problem areas are not necessarily listed in order of importance.

Detrimental Effects of Excess Water

Water possesses detrimental as well as beneficial effects. Detrimental effects of excess water in the North Central Region appear serious and widespread, but these problems have been largely neglected by economists in recent years. Studies in Minnesota and Iowa suggest the nature and extent of these detrimental effects in terms of potential increases in income resulting largely from increased crop yields, reduced production expenses, and timeliness of farming operations.

Results of a 1981 drainage study in southern Minnesota estimated annual net returns to crop production from improved drainage of $42.00 to $147.00 per acre. The study concluded (11), "If landowners want to recover their drainage costs with increases in production, they can spend no more than the present values of these annual returns to drain land. Present values were shown to range from $284.00 to $1,639 (per acre) depending on assumptions regarding discount rate and study area."

According to the results of a recent drainage study (15) in northern Iowa, "If present yield potentials of corn and soybeans are compared with potential maximum yields (high drainage situation) only about 71 percent of the maximum yield potential is currently realized. Applying 1980 price and cost data, the analysis showed that it is feasible to develop further drainage to the extent of 99 percent of the maximum yield potential...."

Potential increases in production from drainage improvements could not only lower per unit costs of production in the drainage areas but could relieve and replace fragile lands subject to water and wind erosion. Most of the land with drainage potential represents more level land less subject to soil erosion caused by water.

Thus, further drainage potential appears to exist in areas studied. This potential invites serious consideration of similar research in other areas throughout the region.

Increasing Scarcities of Water

During the early decades of the 21st century, water problems in the United States may well constitute a greater problem than energy does today. The major difference between water and energy problems is that there are no known physical substitutes for water in satisfying direct and indirect water demands, but there are many known substitutes for petroleum in producing energy. This means that we must learn to live from and with current water supply endowments, subject to some possible amendments discussed later.

Similarities between the current energy crisis and the expected water crisis emphasize increasing scarcities and costs. Water is a necessity of life; it constitutes an essential resource in most economic activities. Thus, increasing costs and scarcities of water are likely to have profound effects upon economic progress, affecting production, employment, income distribution, investment, and debt retirement in the affected regions.

As the major aquifers in the High Plains and Southwest become exhausted, agricultural production is likely to shift from those states to the North Central States, where water is more plentiful. Also, agricultural production in the High Plains and Southwest, with the exception of certain high-value specialty crops, will not likely be able to compete with industrial and urban demands for a diminishing supply of water. Consequently, most agricultural activities will be squeezed out by diminishing water supplies and shifted elsewhere.

Such a shift in agricultural and other production and processing activities to the North Central States, along with increased per capita consumption of water and population growth, may well increase water demand in the North Central States. This increased demand could well strain the water supplies provided by major aquifers in these states, aggravating water shortages, particularly during years when precipitation is short.

Deteriorating Water Quality

About three-fourths of the world's area is covered by water. This is augmented by precipitation and groundwater on and under the remaining one-fourth of the earth's surface. What then is the basis for future concerns about water, particularly concerns beyond water quantity? With problems of water excesses as well as water shortages occurring periodically in the North Central Region, why should the North Central States be concerned about water supplies?

The answer was implied in the words of Coleridge's Ancient Mariner, who, while dying from thirst, lamented, "Water, water everywhere but not a drop to drink." This answer concerns water quality (*20*). The Ancient Mariner was served well by the transportation service of the ocean water that carried his ship, but the same water did not possess the quality to quench his thirst.[1]

Irving Fox (*5*) reminds us, "In the minds of many people, the existing and potential degradation of water quality is our foremost water problem." This problem is magnified by the many and increasing uses for water and their vastly different water quality requirements. The solution to water quality problems rests with water quality management. This solution provides opportunity for helping to avoid or to live with water quality problems in the future.

Traditionally, water, as well as air and soil, has been used to assimilate, dilute, and recycle the residual wastes of human activity. But there are limits to the capacity of water to assimilate, dilute, and recycle all of our garbage. Currently, these limits are being violated through uses of technologies associated with production, fabrication, distribution, and consumption of materials. Presently, use of technologies affecting water quality exceeds the ability to maintain water quality.

Historically, natural resource scarcity has been interpreted in measures of quantities of resources, i.e., gallons of water, depth of soil, barrels of oil. Increasingly, however, people are realizing that scarcity of water and other resources is largely a function of quality.

Quantity of water, for example, may be abundant or even superfluous, but there may not be available sufficient water of a particular quality to satisfy a specific use or demand. The water may be too salty, as was the case with the Ancient Mariner, or too hot, too toxic, etc., for a particular use. As a consequence, a use process may be made more costly, a use may be diminished, or a use may be precluded entirely because requisite quality is lacking, even though there is an abundant quantity of water in the aggregate.

As state and federal governments take action in water quality management, costs of quality improvement meet resistance. As costs of pollution controls press on producers, as prices of products reflecting pollution control costs press on consumers, as pollution control taxes press on taxpayers, and as pollution control measures restrict individual freedom in resource use, voluntary support and enthusiasm for water quality improvement diminishes.

Such resistances may thwart quality improvement unless facts are ascertained and made available to people regarding (1) enacted and proposed water quality standards, (2) costs of achieving these standards, (3) benefits from quality improvements, (4) incidences of costs and benefits in terms of who pays them and who receives them in both the short and

---

[1]For a further discussion of water quality problems, see reference 20.

long term, and (5) nature and effects of antipollution regulations and controls upon individual freedom and choice.

These issues are being decided in legislative, executive, and judicial processes of government. However, under this nation's form of government, support for, and enforcement of, these decisions rests with the general citizenry. Support and compliance, in turn, depend upon how well citizens are informed regarding these important yet complicated issues. How well people are informed, in turn, depends upon availability of relevant information from research and how well this knowledge is extended to citizens.

Detrimental Soil and Water Interrelationships

Complementary relationships between soil and water exist because soil requires water to be productive, as water requires soil. If soil does not receive sufficient water, the crop fails or is reduced. With the possible exception of hydroponics, water requires soil to be productive. However, adverse relationships between soil and water through their joint use may well reduce the productivity of either or both resources in the production process. As indicated earlier, water may erode the soil or reduce the soil's productivity through excessive or insufficient amounts. Also, water may pollute the soil with salts and other compounds that water carries into the soil. On the other hand, soil may harm the water with sediment and with pollutants transported with the sediments or in solution. Adverse effects of soil and water on one another can seriously affect the productivity of each. This example illustrates the interrelationships between resources used in the production process. It also illustrates the limitations and necessary qualification in researching water or any other resource in isolation.

Externalities

One user of water may be able to retain the benefits of use while shifting costs to other users by lowering water quality or quantity. If that user had to bear the costs, there would be some motivation to use the water consistent with the quality and quantity demands of other users.

On the other hand, a water user may be in such a position that, if an outlay is made to maintain or improve water quality or quantity, the benefits from the outlay that shift to other users could not be captured by the user. If such benefits could be captured, the user would be motivated to make outlays that would maintain or improve the quality of water after it leaves that use. Such terms as "side effects," "spillovers," "fallout," or "free-rider," termed externalities by economists, have been applied to such shifts of costs and benefits.

The rationale for these externalities is that the consequences of the actions are external to the individual firm or industry responsible for the actions. Externalities are classified as economies and diseconomies. Beneficial effects are called external economies, and harmful effects are called external diseconomies. Both have in common one phenomenon: The effects shift beyond the user that causes them. The reasons for this shift may be spatial, structural, or temporal.

Although the problem of external economies is important, external diseconomies appear far more important in water quality management. For example, wastes from manufacturing or from chemical fertilizers, pesticides, and livestock moving into streams, lakes, or aquifers may foreclose other uses entirely or make other uses more expensive to undertake. Or they may endanger life and health of human beings.

Allen Kneese (10) concluded, "...a society that allows waste dischargers to neglect the offsite costs of waste disposal will not only devote too few resources to the treatment of waste but will also produce too much waste in view of the damage it causes."

Lack of Water Supply Development

Most precipitation in the North Central Region runs off or evaporates. Most dams and reservoirs in the region are for flood control. Recreational components have been added more frequently only in recent years. Water rights systems are directed mainly toward the uses of water and not its development. These systems are essentially rationing systems that distribute natural water without particular regard to its development.

As future water demands press upon supplies in particular areas and at particular times, increasing consideration will likely be given to water storage, both surface and underground, as well as water conservation. These developments should be anticipated with exploratory research on water supply development.

Exhaustible and Renewable Characteristics of Water Sources

Uses of water draw upon exhaustible, nonrenewable water supplies in aquifers as well as water supplies that are renewed by precipitation or otherwise recharged.

Water and soil are similar in their states of occurrence. Both resources occur in renewable and nonrenewable states. For example, soil that is 20 inches deep, underlain by bedrock, is nonrenewable. It ceases to be productive when the 20 inches erodes away, just as the aquifer containing a fixed amount of water is nonrenewable and ceases to be productive when all the water is drawn out. Of course, each of the resources may become nonproductive economically long before all the soil erodes away or before all the water is drawn out of the aquifer.

Water becomes depleted economically when the cost of production increases to the point that none of the product is demanded from the resource. Thus, as water availability drops through continued extraction, pumping cost increases (from increasing prices for gas, electricity, and other energy sources) deplete the economic supply of water.[2] Thus, the aquifer must be recharged, if possible, or the economic activities must either acquire alternative water sources or relocate in regions with more adequate water resources.

This situation may apply to supplemental irrigation required periodically in the more humid states as well as in western states, which currently rely on irrigation for crop production.

Renewable sources of water in the form of precipitation provide water in time installments. The regularity and certainty of the installments constitute the problem that can be alleviated by supplemental irrigation from aquifers or from stored water.

Allocation of Water Among Competing Uses and Users

Water misallocations are largely a function of inefficient institutions or the absence of needed institutions.

Groundwater is a common property resource that migrates. In many jurisdictions, migratory waters adhere to the law of capture. Consequently, there is little incentive for a particular user to allocate the water as a scarce resource. This leads to water mining.

Since water markets are limited and imperfect, particularly in regard to agricultural use of water, there remains the tendency and incentive to use water as a free good. This may lead to an overuse of water for one use relative to the water's productive value in alternative uses.

Market prices, as scarcity indicators for allocating exhaustible and renewable as well as nonrenewable resources over time, appear unreliable because they underestimate future costs stemming from shortages and reduced management options.

Conflicts Between States and Countries

Serious conflicts are emerging between states and possibly with Canada over water acquisition and use. Research is needed to anticipate, analyze, and solve these conflicts. Economists can contribute much to this effort.

[2]Lee, Jun C., Cameron Short, and Earl O. Heady. 1981. "Optimal Groundwater Mining in the Agalla Aquifer: Estimation of Economic Loss and Excessive Depletion Due to Commonality." Paper presented at the annual meeting of the American Agricultural Economics Association, July 28.

Energy Transportation Systems, Inc. (ETSI) has negotiated with South Dakota to purchase Missouri River water and to transport that water 288 miles through a pipeline to Gillette, Wyoming (21). At Gillette, the water would be mixed with coal pulverized to the size of sugar granules and transported through another pipeline to Oklahoma, Louisiana, and Arkansas. ETSI would be allowed to take 50,000 acre-feet of water a year for 50 years from the Oahe Reservoir on the Missouri River in South Dakota.

Iowa, Missouri, and Nebraska have joined together in bringing suit to stop South Dakota from selling Missouri River water to a private company. The three states fear that a reduction in flow could harm barge traffic, hydroelectric power, fish and wildlife habitats, and recreational uses and, of greater importance, a precedent would be set that would result in future sales of water from the Missouri River. South Dakota claims it owns the water that flows through the state and, therefore, has the right to sell the water.

In the words of a Canadian (17), written over a decade ago, "Canada must begin, now, to develop a policy on water export, a policy that acknowledges that we have a surplus but aims to make that surplus work for Canada, not the U.S. ...It will not do to say, as our government has been saying, that we will develop a policy when the time comes. The time is now. If we wait much longer there is a good chance that our development, our prosperity, our sovereignty will disappear one duty day down an American drain."

A study of the Colorado River (3) concluded that "...allocation inefficiency in water use arising from restrictions on the transfer of water rights is not a new theme. However, the doctrine of prior appropriation, by which users acquiring rights earlier in time have seniority in periods of low river flow, generates additional inefficiencies as a consequence of unequal sharing of risk among appropriators."

Major Water Research Needs

The emerging problems in the use and management of water resources provide the basis for major water research needs in the North Central Region over the coming years and decades. The nine problem areas previously identified require contributions from research scholars in several disciplines, including economics.

From the nine water problem areas, six research needs have been articulated: (1) minimizing detrimental effects of water through improved drainage, (2) improving allocation of existing water supplies, (3) managing water quality, (4) internalizing externalities, (5) supplemental irrigation, and (6) increasing water availability.

Minimizing Water Detrimental Effects through Improved Drainage

There remains the continuing probability of relieving detrimental effects of excess water through improved drainage in areas throughout the North Central States. As indicated earlier, recent studies in Iowa and Minnesota reveal that substantial increases in yields and net income could be realized in the Upper Des Moines River Basin. Drainage studies both at the farm level and at drainage district levels are needed. Regional research could be initiated through inventories of drainage districts throughout the region as well as those level Class 1 and Class 2 lands beyond drainage district boundaries that might require drainage. Homogeneous areas within and between states could be sampled and studied in terms of costs and benefits associated with various levels of drainage under different combinations of technologies, prices, costs, discount rates, yields, and other relevant variables.

Improved Allocation of Existing Water Supplies

Water in the North Central Region has been allocated mainly through water rights systems. These systems, particularly the riparian doctrine, tend to value water as a free resource, with little incentive for the possessor of the water right to allocate water on the basis of its productivity. Studies are needed on water allocation alternatives, varying from competitive market pricing to institutional allocations that would facilitate water use on a basis of its productivity.

Market Mechanisms for Pricing. One alternative for water allocation is to create market mechanisms that price water as a product or factor. Subject to restraints on spatial and temporal occurrence, water could be metered and sold by private, government, or quasi-government entities. In this manner, when a particular quality-differentiated supply becomes scarce, price would ration the supply among the highest valued uses, or call forth an increased supply, as determined by price, or a combination of the two.

This alternative would necessarily invalidate some existing claims to water established through the various rights systems. Although the market alternative has been discussed widely in the literature, little research has been devoted to its implementation (*2, 6, 18*). Water permits, as provided by the Iowa Water Allocation Act of 1957, could be sold at auction to the highest bid. Of course, municipal water supplies are metered and sold by municipalities.

Institutional Rationing. Another alternative, widely used for allocating scarce water among competing uses, is institutional rationing. However, such rationing must be based upon criteria other than mere rights of use, as is presently the case, in providing incentives for getting water into uses yielding the highest net value product. Here again, there are means to accomplish this object.

As an economic basis for institutional rationing (allocation) of water, the value productivity of water could be estimated for alternative uses, with varying amounts of water in each use.

In the value productivity approach, water would be priced through a synthetic market of shadow prices (opportunity costs used in lieu of money prices) indicative of imputed values of water in alternative uses. Alternatively, water could be allocated on the basis of relative contributions by uses to state, regional, or national products or incomes through sector-accounting processes. An Arizona study (22) revealed that the manufacturing sector ranked first in terms of personal income generated per acre-foot of water with a $82,301 payoff. However, more than 90 percent of the water was allocated to the agricultural sector, with payoffs of $14 to $80 per acre-foot.

Interpreting the results of this analysis, the researchers concluded that "...if the problem is to obtain maximum economic growth for the state, this water must generate benefits in excess of costs of transporting and distributing it. Since this is not the case, reallocation of available water becomes the preferred solution."

Managing Water Quality

The quantity theory of water emphasized in and perpetuated through the various doctrines of water rights, with few exceptions, has tended to ignore water quality and to treat all water alike (19). Instead of being homogeneous, however, water is heterogeneous in terms of its properties, its technologically permitted uses, and its economically demanded uses.

It becomes helpful, at least from an economic viewpoint, to regard water supply as differentiated in kinds and grades determined by its quality. Thus, supply and demand functions of water are each regarded as consisting of numerous quality oriented segments, each segment characterized by relatively homogeneous quality.

Qualities of water may be affected by human use, or they may be produced in the natural state. One set of qualities within a natural supply of water may satisfy a particular use but may preclude another use. Furthermore, one use of water may leave a residue or an effluent within the water that diminishes or precludes another use, or which increases the cost of subsequent use of the same water.

This would constitute water pollution, which is a supply-related concept. In economic terms, water pollution means a change in a characteristic(s) of a particular water supply, such that additional costs, either monetary or nonmonetary, must be borne by the next use and the next user either through diminishing or precluding the next use or through forcing the next use to absorb more costs in cleaning up the residue left by the initial use or to develop a new source of water supply.

Internalizing Externalities

Following identification of supply-demand water use entities based upon the criteria of determinants as outlined, economists become concerned with optimizing water use, including internalizing the externalities.

In the use of water, external diseconomies may arise among competing users; costs of use in these cases are shifted to other firms in a spatial or temporal incidence. Through this process, disassociation problems arise whereby certain water costs of one use are transferred to other water uses. Therein, subsequent water uses are made more expensive, or other uses are eliminated entirely or partially. For example, the effluent discharged by a private or public firm, such as a municipality or a dairy product processing plant, respectively, would change the quality of water to the extent that fishing or swimming downstream would be prohibited or decreased.

Allen Kneese (10) concluded that "...a society that allows waste dischargers to neglect the offsite costs of waste disposal will not only devote too few resources to the treatment of water but will also produce too much waste in view of the damage it causes." In other words, external diseconomies provide the incentive for firms to increase water pollution, just as external economies motivate firms to stop short of making pollution control investments.

More research is needed in identifying and measuring these externalities as the basis for developing means for internalizing externalities, both economies and diseconomies.

Supplemental Irrigation

Augmentation of natural water supplies during years or periods of deficient rainfall at critical stages of crop growth is a consideration in the more humid parts of the North Central Region. Several studies have indicated that timing of irrigation investment is a crucial factor in determining net benefits (1, 4, 16). If investments in wells and pumping and distribution equipment are made prior to a series of wet years, interest charges could make the investment unprofitable. But, if investments are made immediately prior to a dry year, or series of dry years, the investments tend to be profitable. Of course, investment and its payoff are affected by the size of investment per acre, which in turn is affected by the technology used.

Studies are needed to relate costs and benefits of supplemental irrigation to fixed and operating costs of equipment, prices, and kinds of products and to antecedent moisture and prospective weather conditions.

Increasing Water Availability

Eventually, probably after the turn of the century, the western part of the North Central Region in particular may consider obtaining insurance in the form of increased water supplies, particularly for years and months of insufficient moisture. Several possibilities exist for increasing water supply. They include weather modification, desalinization of sea water, and importation of water from Canada.

Weather Modifications. Weather modification technology remains questionable. Cloud seeding with dry ice and silver iodide may increase precipitation, but major breakthroughs remain to be made (9). Also, if and when weather modification becomes more reliable, incidences of damages as well as benefits remain to be analyzed and allocated.

Desalinization of Sea Water. Several technologies for desalting sea water have been developed, but all remain much too costly to be used in the United States for most use, even in states adjacent to the oceans. In the future, existing technologies may be improved or replaced by new technologies that have the economic potential for being used.

Importing Water. The possibility exists for exporting water from surplus-producing regions to deficit areas, both within the United States and from Canada (17). Canada has more freshwater per capita than any nation in the world, between 20 and 50 percent of all freshwater on earth. Much of this freshwater proceeds unused into the oceans. Therefore, it is little wonder that as early as 1964 an American engineering firm proposed a $100 billion (about 300 billion in 1982 costs) water plan that would shift the course of part of this Canadian water to the United States, where needs for water were developing. The plan was known as the North American Water and Power Alliance (NAWAPA).

Under the initial plan, NAWAPA would divert part of the north-flowing Canadian and Alaskan rivers by pumping the water 1,000 feet up through huge pipelines to the Rocky Mountain trench, a 500-mile long gorge containing the Columbia, Fraser, and Kootenay Rivers. From there, the water would move eastward, across the Canadian prairies to the Great Lakes, as well as southward, across the drylands of the Great Plains to Mexico.

Since NAWAPA, numerous other proposals have been made, but none has gone beyond the discussion stages. With Canada's large renewable supply of usable water, coupled with the growing demand for water in the Great Plains, western states, and the North Central Region, the likelihood of pursuing further the engineering, institutional, and economic possibilities for meshing Canadian water supplies with U.S. water demands appears inevitable.

Some Research Approaches and Procedures

Approaches and procedures in developing and carrying out research on water resources can be viewed in the context of four categories: (1) dependence of economic analysis on data and assistance provided by physical, biological, and engineering scientists, (2) necessity of interdisciplinary teamwork, (3) the hydrologic unit for water management research, and (4) some appropriate methodologies.

Dependence of Economic Analysis

In addition to collaborative efforts with other social scientists, particularly in sociology, law, and political science, economic analysis of water problems and their solutions strategically depends upon data and the participation of scientists in the physical, engineering, and biological sciences. To obtain and prepare data needed in economic analyses, these scientists should become knowledgeable about economic models and the methods used by economists. With this knowledge, scientists in other disciplines can participate more fully in the analysis as scholars and as full members of the research team rather than as purveyors of data only.

Such collegial participation has several advantages. First, the data and interpretations of data, including qualifications, are more appropriately and accurately used. Scientists realize the importance of their data in economic analyses in terms of the results jointly arrived at. Also, these scientists obtain insights into further data needs as a basis for further research in their own disciplines to generate the coefficients and other information required by models. Furthermore, the economist gains a better knowledge of these needs and is in a position to support the needed research in other disciplines.

Through cooperation, the economist learns to appreciate the limitations as well as the contributions from other disciplines.

Necessity of Interdisciplinary Teamwork

Interdependence among research scholars from the relevant disciplines give credence to the necessity of interdisciplinary teamwork. However, obstacles may discourage interdisciplinary research, including departmental and college-level administrative and budgetary elements.

In instances where such obstacles exist, a cross-disciplinary entity may be established under the vice-president for research or some other university-wide administrator. The water resources research institutes established under the 1964 Federal Water Resources Research and Training Act have contributed much to multidisciplinary water research, including research by scientists from different institutions within a state.

The Hydrologic Unit for Water Research

Water problems arise within hydrologic units and likewise must be solved within those units. Therefore, the management of water resources applies mostly within hydrologic units. These units generate the supply and quality of water. Likewise, the hydrologic unit serves as the conduit for eliminating excess water accumulation. Thus, water research should be conducted within this unit to deal with the problems and their remedies.

The hydrologic unit may be a river basin, an aquifer, a lake, or any other water area. It represents a common source of water supply that can be aggregated to a river basin (Missouri River), an aquifer (Ogallala), a lake, or a chain of lakes. The unit may be disaggregated to a watershed, a drainage district, an irrigation district, or any other use-oriented district.

The management unit for water thus naturally conforms to the hydrologic unit of water. This unit provides a commonality of problems, solutions, and management. This unit also provides the research opportunities for contributions by economists.

Some Appropriate Methodologies

Numerous methods have been used to define and to suggest means for achieving an economic optimum in the use of water resources. Linear programming constitutes one of the most commonly used procedures in analyzing water problems and remedies.

In recent years, open Leontief input-output system with linear programming models have been applied at state and regional levels to analyze water uses and allocations among sectors. "In contrast to the general multimarket equilibrium analyses," observed Orris Herfindahl and Allen Kneese (7), "the system contains no utility functions, and consumer demands are taken exogenously. Also, the industry rather than the firm is taken to be the unit of production, and the production function of each industry is a constant-coefficient type so that it presents no optimization problem."

An application of this type of model is illustrated in a study of 12 counties in northwestern Iowa (*14*). Purpose of the application was to investigate water availability for the region's population and economic activities as projected by the state to the year 2020, given the water use rates of 1967 and the production structure embodied in the 1967 input-output table for Iowa. It was assumed that each of the water supply areas is not augmented by transportation of water from other areas and that all water in the region is homogeneous regarding quality. Results of the application suggested that the region possesses sufficient water supplies for the area's projected population and economic growth. When the region was broken down into counties, however, the expansion of irrigation caused water to become a limiting factor to economic growth in several counties, thus necessitating water transfers from surplus areas to areas with shortages.

Another study using goal programming and input-output analyses was applied to the Iowa economy. This study provided guidance in the allocation of water throughout Iowa using input-output analysis and goal programming combined. The 77 sectoral classifications of the U.S. economy were adapted to the state. Eight water supply areas were used. Nonirrigation water requirements were readily accommodated by estimated water availabilities throughout the state; however, water withdrawals in the utilities sector was found to affect water qualities in most areas. Only crop irrigation, the largest estimated demand for water in the state, would press water demands beyond available supplies in all areas of the state except the Missouri and Mississippi water supply areas. But the use of water from these supply areas could conflict with water demands from other states bordering these rivers (*12*).

REFERENCES

1. Babula, Ronald. 1978. *Development of a methodology for analyzing irrigation investment with a case application.* M.S. thesis. Iowa State University, Ames.
2. Barnett, Andy H., Russell D. Shannon, and Bruce Yaudle, Jr. 1974. *Establishing markets for water quality.* Report No. 44. Department of Economics, Clemson University, Clemson, South Carolina.
3. Burness, H. Stuart, and James P. Quirk. 1980. *Water law, water transfers and economic efficiency: The Colorado River.* The Journal of Law and Economics 22(1): 111.
4. Colbert, John. 1978. *Determining methodologies for estimating the value product of water used for irrigation with application to selected cases.* M.S. thesis. Iowa State University, Ames.
5. Fox, Irving K. 1970. *Promising areas for research on institutional design for water resources management.* In Dean T. Massey [editor] *Implementation of Regional Research in Water-Related Problems.* University of Wisconsin, Madison. p. 32.
6. Gaffney, Mason. 1962. *Comparison of market pricing and other means of allocating water resources.* In *Water Law and Policy in the Southeast.* Institute of Law and Government, University of Georgia, Athens. pp. 195-227.
7. Herfindahl, Orris C., and Allen V. Kneese. 1973. *Economic theory of natural resources.* Charles E. Merrill Publishing Co., Columbus, Ohio. p. 54.

8. Howe, Charles W. 1979. *Natural resource economics.* John Wiley & Sons, New York, New York. p. 89.
9. Kerr, Richard A. 1982. *Cloud seeding: One success in 35 years.* Science 217: 519-521.
10. Kneese, Allen V. 1964. *The economics of regional water quality management.* Johns Hopkins Press, Baltimore, Maryland. p. 43.
11. Leitch, Jay A., and Daniel Kerestes. 1981. *Agricultural land drainage costs and returns in Minnesota.* Staff paper series P81-15. Department of Agricultural and Applied Economics, University of Minnesota, St. Paul. p. 36.
12. Mensah, Edward Kingsley. 1980. *The management of water resources: A synthesis of goal programming and input-output analyses with application to the Iowa economy.* Ph.D. dissertation. Iowa State University, Ames.
13. Pagoulatos, Angelos, and John F. Timmons. 1978. *Management options in the use of exhaustible and nonrenewable resources.* Revista Internazional di Scienze Economiche e Commerciali XXV: 504.
14. Rhee, Jeong J. 1980. *The combination of input-output linear programming for water resources management: An application to Northwest Iowa.* Ph.D. dissertation. Iowa State University, Ames.
15. Schult, David L., Thomas E. Fenton, Roy D. Hickman, Howard P. Johnson, Rameshwar S. Kanwar, and John F. Timmons. 1981. *Present and potential agricultural cropland drainage conditions in the Upper Des Moines River Basin.* Iowa State University, Ames, in cooperation with the Soil Conservation Service, U.S. Department of Agriculture, Des Moines, Iowa. p. ii.
16. Soonthornsima, Wipada. 1981. *Economic feasibility studies of irrigation in northwest Iowa.* M.S. thesis. Iowa State University, Ames.
17. Stewart, Walter. 1970. *The sellout that could spell the end of Canada.* McLean's (March): 4-5.
18. Tabor, Eric J. 1980. *Proposal for a regulated market of water rights in Iowa.* Iowa Law Review 65(4): 979.
19. Timmons, John F. 1967. *Economics of water quality.* In *Water Pollution Control and Abatement.* Iowa State University Press, Ames.
20. Timmons, John F. 1980. *The quality of water: Problems, identification and improvement.* In *Western Water Resources: Coming Problems and Policy Alternatives.* Westview Press, Boulder, Colorado.
21. Upper Midwest Council. 1981. *Energy Transportation Systems, Inc., buys Missouri River water.* Upper Midwest Report 11(6): 1.
22. Young, Robert, and William Martin. 1967. *The economics of Arizona's water problem.* The Arizona Review 16(3): 10.

12

Water Resources Research: Potential Contributions by Sociologists

Peter F. Korsching and Peter J. Nowak

Water and its quality, quantity, allocation, and management will be the critical environmental concerns of tomorrow. Even today few areas of the United States are without water-related problems. These problems include inadequate water supply, groundwater depletion, surface water pollution, groundwater contamination, domestic water supply contamination, flooding, erosion and sedimentation, and drainage. In some areas these problems are already severe. Water is a basic necessity for life. Unlike other natural resources, such as fuels and minerals, however, there are no substitutes for water.

It would be a blessing if the solution to water problems were simply physical, biological, or even economic. That would greatly ease the process of seeking and implementing solutions to the problems. However, many technological fixes (even with favorable biological consequences) and economic payoffs (even with high benefit/cost ratios) are not necessarily the best solutions when other factors are considered. In particular, sociological factors often mediate or even negate potential technological and economic benefits. For example, the ability to recognize the problem is not distributed normally across the population, but varies along social, cultural, and ecological dimensions. Other factors include the distribution of skills and resources necessary to take advantage of these ameliorative technologies or programs with financial incentives; the distribution of social and economic consequences across groups in the affected and unaffected populations, as well as the reaction of these groups to the consequences; relative to the affected population and other agencies, the effectiveness of the involved agencies in communicating, educating, and implementing the program; and the nature of political mediation in the distribution of benefits or imposition of costs or restrictions. In policy statements, planning documents, and program outlines, sociological factors such as these are often found in discussions of what constitutes a successful or unsuccessful program.

A good example of the failure of a program to resolve a situation for which the technology exists is the problem of nonpoint-source pollution caused by soil erosion and sedimentation. Technologically speaking, the means to effectively control sedimentation and the impurities transported by the sediment have existed since the 1930s (30). Yet billions of dollars and half a century later, little has been achieved in controlling nonpoint-source pollution. Part of the reason for this failure may be the types of information used in developing the programs (25). The biological effects of sediment and other pollution in streams, lakes, and other bodies of water were largely understood. The physical process of soil movement and the methods to prevent it were largely understood, as was the anticipated national reaction to the application of financial incentives. What evidently was not understood, and as a result not incorporated into the program, was the role of social factors.

This is not surprising. In 1980, the U.S. Department of the Interior's Office of Water Research and Technology requested that each state develop a five-year water resources research plan for the years 1982-1987. Basically, the plans were instituted to identify water problems, research needs, and the priorities associated with these problems. A significant amount of variation exists among states in terms of the methods used to identify problems, needs, priorities, and the scope of the reports. What is particularly enlightening is the one area of consistency across these reports. This is the lack of emphasis given to social concerns. A fairly typical example is the Kansas plan, which covers the social factors with the abbreviated caveat that research should be conducted on the "legal, institutional and social implications..." of alternative water management systems (20). The clincher is that in the recommendation of funds for research nothing is allocated to this priority.

Despite these and other indications to the contrary, the importance of social factors is slowly gaining recognition. Recent reports on soil and water conservation make special mention of the important role of these factors (6, 35). Part of the problem may be that sociologists themselves have not been sufficiently aggressive or adept in demonstrating what their discipline has to offer these problem areas. As a result, it was left to researchers in other disciplines (such as natural resource management) who had just enough sociology in their training to include a few sociological variables along with the psychological, political, and economic variables in their research. The literature in the water resources journals and books is replete with such quasisociological studies. Although they often lack theoretical grounding and methodological sophistication, they cannot be faulted too severely because they have attempted to fill voids in areas largely ignored until recently by sociologists.

Water Problem Awareness and Citizen Participation

One of the fundamental principles of the decision-making process in the solution of a problem is that before any other considerations there must be awareness of the problem. This seems almost too self-evident and simplistic for elaboration, but it is something that public policymakers, public administrators and change-agency personnel often overlook. There is an implicit assumption in many programs that the intended audience of the program has the same type and level of awareness as the policymakers who initiated the program and the change agents who will be implementing it. But this is not the case. An intended program audience may (1) not be aware that there is a problem; (2) be aware that there is a problem in a general sense, but have no feeling of personal involvement in the problem or efficacy in its solution; or (3) be aware that there is a problem in a general sense but question the credibility of the change agency or proposed program in defining and dealing with the problem.

Lack of problem awareness of the second and third types, which is not uncommon, can have serious consequences in implementing a program. An examination of the awareness of soil erosion and resulting sedimentation among Iowa farmers discovered an inverse relationship between awareness of these problems and proximity to the farm operation.[1] Farmers were well aware that the Midwest had soil erosion problems and that Iowa had soil erosion problems. But they expressed less awareness of problems in their own community, and only about half stated that they had a serious soil erosion problem on their own farms. This proximity effect—there is a direct relationship between the perceived severity of the problem and the distance from one's own situation—has also been found in other studies (33). In other words, it is usually someone else—the farmer farther down the stream or in another community—who is causing the erosion problems. More importantly, these perceptions are not very accurate. It was determined that 71 percent of those individuals who claimed that there was not an erosion problem on their farm actually had erosion rates exceeding the accepted limit of five tons per acre per year (33).

Another example is a study in 12 Iowa communities on the effective-

[1]Korsching, Peter F. 1981. "Some Iowa Soil Conservation Problems and Prospects." Paper presented at the Land Use/Ownership Conference, Rural Iowa, Inc., Des Moines.

ness of mandatory versus voluntary water conservation policies during a drought (27). Although mandatory policies proved more effective in general, in two communities with voluntary programs consumption of water was also reduced significantly. In these two communities the credibility for water shortages was high because nearby communities were known to have serious water shortage problems. "The key to successful conservation seems to be the credibility of the water shortage rather than the penalties per se" (27).

Awareness, then, can be critical to the effectiveness of any program developed to solve water resource problems. Awareness is the first step in acceptance of a new idea (40) and in the implementation of a planned change program (56). Many resource managers are learning that the real problem they face is one of creating an awareness among their clients of resource problems, recognition of existing practical solutions, and knowledge of available technical and economic assistance.

Creating Awareness

One technique for creating awareness is through public participation in the decision-making process. However, public involvement in decisions—even though the decisions will have a direct impact on the public—has had mixed acceptance by those traditionally involved in the decision-making process. A number of justifications, both pro and con, have been put forth relative to the use of public participation. Among the reasons usually cited in favor of public participation are that it develops within the individual a sense of control over the life situation, it legitimizes the development process, and it can actually lead to better decisions about water policy by incorporating broader perspectives than exist within the agency (10). Those against public participation contend that it unduly constrains the agency in its work, that water resource policy is a highly technical matter upon which only experts can make intelligent decisions, and that biases occur anyway through selective participation (10).

Whether pro or con in their leanings toward citizen participation, most public decision-makers fail to realize the opportunity for creating awareness of the problem through this technique. Ideally, citizen participation should be a two-way communication in which the agency communicates to the public its perceptions of the problem and viable solutions, and the public responds with its own perceptions, thus becoming involved in the decision-making. Unfortunately, little emphasis in research has been placed upon the agency-to-public communication/information/education linkage that would help to better understand how to use that linkage effectively in creating awareness. A recent book on public participation in water policy decision-making (10) contains only a brief comment on the need for educational efforts to ensure an informed, participating public (*51*). Most of the book is devoted to when, why, and how information from the public could and should be used in making decisions.

This is not to deprecate the traditional reasons for the use of citizen participation in making policy or developing programs or projects. However, because it is inconceivable that decisions in matters of water use and policy development would not affect the public, citizen participation should be incorporated *a priori* into any plan. Furthermore, to develop real citizen participation and to avoid tokenism or cooptation (*2*), the agency should develop a situation in which participants are informed about the problem or issue under consideration and, therefore, have a meaningful input into the decision-making process. Citizen participation thus will be an awareness-creating technique.

There are some issues that need to be resolved in this area for which there is currently little data available to provide any clear direction. Several citizen participation techniques are available. These range from direct participation techniques, such as open planning meetings, to indirect techniques, such as citizen advisory committees and need assessment surveys. The types of communication linkages or mechanisms used in creating awareness depend upon the nature of the change and how it will affect the target population, the characteristics of the target population, and the characteristics of the change agent as perceived by the target. The change agent must be perceived as reliable, creditable, and trustworthy for efficacious communication to occur (*56*). The more serious the effect of the change upon the target population and the greater the opposition or diversity of opinion, the greater the need for more personal and closely monitored communication strategies. Research is needed to determine the optimum type of involvement and the extent of involvement for particular types of issues. Citizen involvement in resolving issues can be expensive and time-consuming. Yet in the long run the failure to involve the public may be even more costly in time and money (*23*). In any program of planned change, there invariably are differing and opposing positions between the target population and the change agency. These differences may even exist among various segments of the target population. What types or combinations of strategies would best enhance awareness of the issue and empathy with the change agency's position? What educational techniques are most conducive to incorporation into the participation strategies? These are questions for which satisfactory answers do not exist.

Using Citizen Participation Input

In relation to citizen participation *per se*, there are some important questions about how to use information obtained through participation in the planning process. The inability to link a multitude of constituencies in an evolving design and development process constitutes one of the major barriers to the design of alternatives (*21*). Although this is problematic, it is only through linking a multitude of constituencies that real citizen participation can occur. Roland Warren (*50*) contends that in the real political world there is no single "public interest." In water resource issues there are usually many publics, both interested and affected, with differing opinions (*51, 54*).

A number of studies have examined the issues of measuring public opinion, analyzing the data, and using the results in policy formation (*7, 31, 41, 42*). Unfortunately, much of the work, especially in the use of the data, has tended not to be cumulative. Much more research is needed in the development and evaluation of models that will assist decision-makers working with complex water resource problems and large and diversified publics to collect theoretically relevant information and to use that information in the development of policy that will benefit the broadest range of constituencies.

Sociologists have part of the solution to this problem in the community need assessment techniques that have been developed and refined in recent years (*3, 24*). The problems addressed with these techniques, however, have been fairly general and simple, and the analyses of the information collected have been fairly simplistic, at least in terms of policy recommendations. Analytic models must be developed that can incorporate large-scale surveys and other types of data and yield information useful for making decisions on complex issues.

Allocation of Water

Although the United States has a sufficient supply of water for the present and near future, water is often inappropriately distributed, both geographically and temporally. This can and does cause regional and seasonal water shortages (*11, 49*). The necessity for temporal allocation of water through storage in wet seasons for use in dry seasons, or regional allocation through transfer from areas with surplus supply to areas with a deficit, is already occurring. In areas with a dwindling supply or shortages, there will be a growing need for reduced allocation or reallocation among competing sources.

Irrigation development in the Columbia River Basin is on the verge of necessitating reduced allocation or reallocation (*53*). Development of center pivot technology has allowed new land to be brought into agricultural production (*32*). Irrigation has also expanded because developers do not experience the full cost of development, obtaining both water for irrigation and electricity to operate the pumps at prices that do not reflect actual costs (*16, 53*). Other sectors of society are bearing the cost. The emerging problem is that increased irrigation is reducing streamflow to the point of reducing hydroelectric power production. Without delving into a complicated discussion of costs, benefits, and trade-offs, decisions must be made on allocation of water between agricultural users, which benefits one particular sector, and the production of inexpensive hydroelectric power, which benefits other sectors.

But making and executing such a decision is a complex problem. Some dimensions of such decisions are these: Who makes the decisions? Who will be affected? What will the effects be? Will those harmed by the decisions receive just compensation? How will compensation be funded? Involved in all these decisions are questions of equity—who benefits and who is harmed? In water allocation or reallocation, the social effects and the issue of equity are of interest to sociologists.

Allocation Mechanisms

In the allocation of water, paramount to all other factors are the rights to water through the legal statutes of each state. Basically, there are two types of legal systems controlling water rights, the riparian doctrine and the prior appropriation doctrine, although some states use a combination of the two (52). Prior appropriation means that the person who uses the water first will always have senior right to it, with diminishing rights for consecutive users. Riparianism means that each owner of land along a stream is entitled to reasonable use. Rights to underground water depend upon whether the water is in a stream or is percolating water. The same statutes that apply to surface streams generally apply to underground streams. For percolating water, there are five doctrines ranging from absolute privilege to correlative rights. As one would expect, prior appropriation doctrines are most common in the dryer, western states. Riparian doctrines are most common in the more humid Midwest and East.

Legal battles over the allocation and use of water have been a continuous phenomenon since the establishment of water laws. Many states are developing new legal and institutional arrangements, however, to better manage problems of decreasing water supplies, water shortages, and water pollution (17). Also, many states are developing comprehensive state water plans that have a strong possibility of achieving regulatory status and thus becoming an important tool in water resource management (17). Along these lines, Nebraska has instituted the Ground Water Management Act to address increasing concern about water mining resulting from center pivot irrigation (1). The act authorizes a number of mechanisms for controlling the mining of groundwater. These include well spacing, well drilling moratoria, pumping rotation, and quantity limitations.

In some situations, the water rights doctrines are being replaced by the public trust doctrine. The public trust doctrine places a limit on private property rights to protect and strengthen social rights through such legal mechanisms as eminent domain (36).

A current debate revolves around what mechanisms should be used to allocate water. In opposition to institutional mechanisms, such as regulations with administrative and enforcement structures, and legal mechanisms, such as eminent domain, some experts suggest that market mechanisms might more effectively allocate water (34, 46). Through market mechanisms, water would be put to its highest and best use in economic terms. Arbitrariness and nonrationality would be removed from water allocation. For instance, Gordon Sloggett (44) suggests that groundwater depletion may eventually correct itself, owing, among other factors, to higher pumping costs. Others question the advisability of such a system. Timothy Tregarthen (45) states that "best use" is a fable; "...the notion 'best' rests on each individual's perception of his or her own welfare." Finally, some aspects of water, such as a beautiful stream, are simply impossible to quantify in economic terms.

Equity in Allocation

Issues relating to water allocation obviously are complex. There are value and ethical implications in all allocation mechanisms and decisions that cannot be circumvented. Controlling groundwater mining in Nebraska is a case in point. There is the value question of whether the use of that water for irrigation is in the best public interest and meets broader societal goals. Allowing the market mechanism to make adjustments through increasing costs of withdrawal will eventually mean radical changes in the nature of agriculture and the rural communities when water withdrawl becomes too expensive for irrigation (44). Also, several mechanisms authorized by the Ground Water Management Act for controlling groundwater depletion, such as well spacing and drilling moratoria, are of questionable equity. They favor those who have been causing the problem (1).

Similarly, in the Columbia River Basin, developers of pivot irrigation agriculture are gaining a disproportionate share of the benefits and not paying the costs (32). They are obtaining cheap water, pumped with inexpensive, industrial-rate electricity, while simultaneously reducing streamflow. This reduces hydroelectric power generating capacity, forcing power companies to switch to more expensive thermoelectric generation.

There is another indirect effect with important consequences also. Center pivot technology for irrigation requires high investment and large blocks of land. This encourages the establishment of corporate farms, reduces the farm labor force, and jeopardizes the viability of small service centers (32).

To resolve some of these problems, research is needed, first of all to determine the long-range social and institutional effects of the current systems of allocation. What are the effects of the development and growth of center pivot technology upon the structure of agriculture and surrounding agricultural community? Also, if the users of the resources are not paying their fair share, research is needed to determine how the costs are distributed and what the specific and overall social burdens are for supporting this system.

Use, control and reallocation of water resources undoubtedly will become more common in the future. Some institutional and legal regulatory mechanisms for using and reallocating water have been discussed. Little is known, however, about the effectiveness and efficiency of the various mechanisms available or being developed. The problem is complicated by the number of organizations with different jurisdictions that are involved in this issue (52). The body of knowledge and theoretical perspectives of sociology on complex organizations and interorganizational relationships can be used to increase understanding of organizational roles in the process. With additional research, it might be possible to increase the efficacy of the institutional mechanisms by improving interorganizational coordination and cooperation, the development of new organizations and institutions, and improvement of relationships with the clientele or affected population. The sociology of law can also play a major role by providing a framework for examining the socio-political process of the development, institutionalization, and enforcement of legal systems and in assessing the impact, ethics, and efficacy of various legal mechanisms for controlling water use.

Social Impact of Water Policies

Whether in the area of water allocation or another area relating to water resources, such as pollution abatement, flood control, wetlands preservation, or water augmentation, the policies developed to contend with the problems and issues will have a variable impact on different societal groups. Some groups in society will derive more benefit from a particular policy than other groups, and some groups will bear more of the cost than other groups. Techniques are available for measuring the aggregate economic costs and benefits for water and other natural resource policies, even, to a degree, the costs to specific societal groups (5, 8).

In addition to economic benefits and costs, any policy will also include social costs and benefits, which will also be unevenly distributed. The measurement of these costs, however, has not been as well developed as the measurement of economic costs. These costs are not reflected in the marketplace because they cannot be assessed by a monetary value; measurement of the social impact of water policies is thus elusive (12).

This is not to say that social impacts cannot be measured. Sociology has an extensive literature in social impact assessment (28). A distinction must be maintained, however, between the impact of water policy and the impact of some specific water project. Water policies are general systems of legal regulations. The systems include plans for implementation and administration of policies within some defined population or geographic area. Policies usually reflect the predominant value system of the larger society. They are formulated to maximize societal goals. Not only do they affect the distribution of benefits and costs in the present and future, they also may restrict the benefits and increase the costs in the present to ensure the existence of the benefits in the future. As such, policies are futuristic; their developers have visionary concepts of what the future could and should be and what long-range directions are needed to attain those goals.

Within the context of water policy are specific actions to achieve policy goals. A specific action may be the construction of an impoundment or the interbasin transfer of water. These specific actions are projects. They are more limited than policy in their impact on a geographic area and population. Also included are actions that may not be covered by an existing policy but that may create problems requiring a policy response, such as groundwater mining. Social impact assessment is most effective for such actions of more limited scope.

Several techniques have been used or proposed for assessing the impacts of water policy. One is a ground-up community analysis for assessing the social effects of water quality management programs (56). Ground-up community analysis involves the selection of several neighborhoods from areas with differing management programs. Intensive analysis of each neighborhood is then conducted and comparisons across neighborhoods are made to assess the differing effects.

On a larger scale is an analytic framework for defining social well-being that addresses the problem of not being able to measure social well-being in monetary terms (12). The problem with measuring social well-being is this: With several parties and several alternatives from which to choose, it is impossible to develop a criterion that will yield stable results relative to the public's maximal or optimal welfare. The solution is not to rely upon an expressed preference, but rather to place the choice in a context—the context of futures foregone, i.e., the degree to which a choice limits or eliminates future options.

The futures foregone technique uses a panel of judges with expertise in the resource problem to evaluate policy alternatives. This panel determines the degree to which the alternative will restrict future options for relevant sectors of the affected societal group. Use of the technique to evaluate resource management alternatives in a planning unit of the Forest Service showed its potential in resource planning.

The futures foregone technique is simple, short, and relatively inexpensive. Sociologists, however, might have some concern about its elitest approach—the small panel of expert judges—for making decisions about social welfare.

Another system, CODINVOLVE, involves a much broader input of data (7). The input comes from affected citizens in response to specific policy questions through various media, such as personal letters, response forms, workshops, petitions, reports, and other methods. All information obtained is coded, tabulated, and analyzed. In using the technique to examine four classification schemes for a river, the Forest Service received and analyzed 6,805 inputs. CODINVOLVE is thus more comprehensive, but also more expensive.

Other schemes could be reviewed, but that would add little to this discussion. As David Freeman (12) stated, "...the toolbox labelled 'Social Well-Being' has remained notably empty." This is particularly true for social impact assessment of water resource policy. Some problems with the various techniques proposed are (1) a lack of grounding in sociological theory to make the results meaningful and compatibile with a larger knowledge framework, (2) lack of application and evaluation in comparative and longitudinal situations to assess the predictive power of the techniques relative to impact, and (3) independence of the developers of these techniques, which has led to the development of unique and idiosyncratic techniques that have few potential applications in real situations.

Given this state of affairs, sociologists have a major role to play in developing viable techniques to assess the social impact of water policies (37). They have only begun to scratch the surface, however. Research should begin with an examination of existing techniques. Useful concepts, theories, and methods should be culled and synthesized. Once techniques are developed, they should be applied, evaluated, and revised as necessary. No one technique is likely to suffice. The optimum impact assessment probably will come through some combination of techniques. For policies affecting large geographic areas with diverse populations, this may involve modeling techniques that use available secondary data for the region as a whole, combined with primary data collected for specific sample areas within the region. Fruitful areas from which researchers might glean information in the development of techniques and models include social indicators research, social impact analysis, and evaluation research.

Water Policy as a Social Planning Tool

In discussing the social impact of water policy, something should be said about the use of water policy to achieve goals that do not relate directly to water use. For instance, manipulation of water resources could be a valuable tool in land use control (*52*). However, one study showed that water resources policy had little effect on controlling or channelling urban development (*39*). The costs of water-related facilities and services are relatively small, at least at present, when compared with the cost of other essential community facilities. Furthermore, federal policy on water-related problems, such as water pollution, is likely to solve problems only where the policy is compatible with local growth strategies (*29*). Thus, water plays a minor role in whether a community will grow or decline. The adoption of higher environmental standards may provide increased leverage to water policy as a tool for controlling urban development. This is important to the farming sector, which is losing prime farmland to urban sprawl.

One of the best examples of a policy's failure to achieve an indirect goal was the Reclamation Act of 1902 (*19, 22, 26*). This act limited the purchase of water by an individual farmer from federally funded irrigation projects to that which could be used on 160 acres. It also stipulated that the farm owner reside on the farm for the farm to be eligible for irrigation water through Bureau of Reclamation water projects. It is conspicuous that, although the nation supports the idea of the small family farm (*47*), the act has seen little enforcement over the years. Although most attention to this issue has centered in California, specifically the Imperial Valley, there are possible impacts in other areas as well. Some areas in the North Central Region, for example, have irrigation water provided by Bureau of Reclamation projects.

The nonimpact of policy is an interesting area for research. If a policy does not have the desired effect, societal values and goals expressed by that policy are being subverted. For instance, failure to enforce the Reclamation Act of 1902 has had profound results on agriculture's structure and the social lives of rural community residents that run counter to the values and goals expressed by the policy (*13*). The current spread of center pivot irrigation may have equally profound effects and should be monitored by researchers (*32*). Unfortunately, these nonimpacts and unanticipated consequences are not as easy to identify as the direct impacts. Researchers must be attuned to current affairs, with insight guided not only by sociological theory but also by intuition and imagination.

Potential Acceptance of Different Water Policies

Formulating policies for the development and management of water resources may be much more simple than implementing those policies. To a large degree, successful implementation of policies depends upon the perceptions of affected groups about their role in the problem and the social and economic benefits or costs they must incur in the solution.

Of equal importance is how policy goals mesh with their own goals and values, i.e., the degree to which they are committed to the purposes and goals of the policy.

Roy Rickson (38) commented, "No public agency can function without legitimation of its programs and activities. Government programs must in some sense be supported, accepted or at least tolerated by significant segments of the public if they are to have any chance of reaching their objectives."

To date, little research has been done in this area. Policymaking has largely been the purview of public decision-makers, with the major considerations being the results of technical and economic analysis of the problem and the political feasibility of the recommended solution.

Because of the nation's overriding value systems, only certain approaches to the solution of problems are considered politically feasible. These approaches emphasize self-help, local initiative, citizen participation, and the public interest. Within these strictures, there are but two legitimate roles for public implementing agencies: educator and facilitator. The agency can educate the client about the problem and alternative solutions to the problem, or the agency can provide the client with resources, such as financial aid or technical assistance, that the client does not have but which are necessary to solve the problem. Guiding these policies are assumptions that the client is rational and will respond appropriately to the information and to the technical and financial assistance provided by the implementing agency. As can be seen in the case of traditional soil and water conservation programs, however, these policies lack effectiveness.

It is not that the public necessarily objects to government intervention in development and management of water resources. Indeed, recent surveys indicate that the public strongly approves of government involvement in such areas as soil and water conservation (15). But what type of intervention, as determined by the type of policy, is acceptable? One of the few studies to examine this issue closely found that, in relation to soil and water conservation, farmers were more likely to approve of policies that did not force compliance (25). They favored policies of economic incentives and educational programs over economic penalties and legal regulations.

Another study examined the opinions of nonfarmers toward differing water resource conservation policies (9). The sample included public officials involved in water policy and administration in local, state, and federal offices; private developmental organizations; private protectionist organizations; and operators of businesses, both users and nonusers of water in their operations. The four policies examined were (1) positive incentives (e.g., allow farmers to sell saved water or use water for purposes not on permit), (2) voluntary compliance, (3) negative incentives (e.g., reduce water allocation or charge higher fees for those in violation of reasonable use), and (4) water use monitoring. Local public officials, those having closer contact with the public, gave the greatest support to positive incentives as a policy to promote water conservation and the least support to voluntary compliance, negative incentives, and monitoring. State and federal officials, those being somewhat more removed from contact with the public, gave the greatest support to voluntary compliance, negative incentives, and monitoring.

As one might expect, members of private, developmental organizations were more in favor of the self-help-with-assistance types of policies, voluntary compliance, and positive incentives, as opposed to members of protectionist organizations, who were more in favor of negative incentives and monitoring. Finally, operators of businesses dependent upon the use of water were less supportive of all types of conservation programs than operators of businesses not dependent upon water use.

What does this research mean? It means that generally people prefer policies that allow them to make their own decisions and that offer them assistance when needed. Unfortunately, these policies have not been overly successful (27). But if mandatory policies prove more effective, there are also additional costs involved. These costs involve surveillance and the application of sanctions when necessary (56). There may also be social costs, such as alienating particular sectors of the public, losing support for other programs administered by the agency, and creating negative attitudes toward future conservation efforts.

Research is needed to better understand the acceptability and the effectiveness of various types of policies for developing and managing water resources. This will become increasingly important as problems reach serious and critical levels. Obviously, the preference would be for noncompulsory policies if they could be made effective, or compulsory policies that are at least palatable to a majority of the population concerned. What is the threshold of control beyond which the policy is perceived as being repressive? That is the critical question. Are there compulsory measures that offer a high level of control but require a low level of surveillance and use of sanctions?

One way to achieve high control with little surveillance is to develop a commitment to the policy by those affected. In other words, those affected by the policy should themselves have an investment in the policy to insure its success. A good example of such a policy is the Nebraska Ground Water Management Act, which allows groundwater users to "...administratively impose ground water controls on themselves through local multi-purpose Natural Resource Districts" (1). Nebraska, like many other western states, has had an increase in groundwater withdrawal for irrigation leading to groundwater mining. If groundwater mining remains unchecked, it could lead to depletion of the groundwater, which will have negative effects upon the local and regional economies. It is actually to the benefit of groundwater users to control groundwater withdrawal. However, "...successful regulation of groundwater mining is not widespread in the western states, primarily because irrigators incorrectly assume that groundwater regulation threatens rather than enhances their economic interest" (1). It would seem, then, by demonstrating to groundwater users that regulation is to their benefit and by establishing the institutional means for self-regulation, compulsory policies can work.

Sociological research should be conducted on Nebraska's Natural Resource Districts. This research should attempt to determine the basic sociological principles upon which these organizations are built and examine the effectiveness of the organizations and the factors that contribute to or detract from their effectiveness.

Social Impact of Water Development

With increasing demand upon available water, it will be necessary in the future to stabilize water supplies in areas with high-intensity use. A number of methods and schemes have been proposed to assure the availability of water. The most feasible methods for the immediate future are construction of additional impoundments, interbasin transfers, and deep aquifer supplies (11). Where major environmental modifications occur because of water supply development, there are bound to be concommitant impacts upon the population in the immediate area of the project. This can mean changes in land use, changes in business and industrial patterns, disruption of the social and cultural organization of communities, and disruption of individual and family lives.

The consequences of such planned social changes have been the subject of more sociological research and documentation than any other topic relating to water resources. In the late 1960s and early 1970s, investigations were conducted in Kentucky, Ohio, Utah, and several other states to understand the social impacts of water resource development. These studies included such topics as attitudes toward water resource development, public participation in the decision-making process, effects of forced migration due to land acquisition and property condemnation procedures, and impacts on communities before, during, and after construction (18). More recent studies focused upon such topics as differences in goals and values of the change agency and the affected population. One of the most comprehensive was a four-year study of the construction of the Lake Shelbyville Reservoir in southern Illinois. This research involved a multidisciplinary team to examine the physical, biological, economic, social, and political effects of the reservoir's construction (4).

Results of these numerous investigations are fairly cumulative, allowing generalizations about disruption to individual, family, and community lives; relationship of the change agency (in these cases usually the
Corps of Engineers) with the population; and public participation, or
lack thereof, throughout the process.

Community Impact

Despite the work done in this area, more is needed, especially research
that examines the impact on the larger community. There is some question about who benefits from these projects. Is it the affected area,
usually a rural area, that receives the benefits of development (flood control, recreation, water supply)? Or is it some other area, usually an urban
area somewhere downstream that accrues most of the benefits? Research
on the impact of reservoir construction in the Pacific Northwest found
that there are both costs and benefits for communities in the vicinity of
the construction site (14). These costs and benefits may not be equally
distributed across different sectors of the population. The business sector
usually accrues short-term benefits through greater business activity. The
poor and the aged, especially if they are among the displaced population,
will suffer the greatest social and economic costs. The entire community
will bear certain costs through greater demands placed upon community
services and facilities. Other social costs include the creation or attenuation of social problems through the influx of a population that is not
integrated into the community system.

On a larger scale, Bernard Shanks (43) examined the impact of the
massive Missouri River water developments in Montana and North Dakota. Construction of five large dams in this region was supposed to
yield economic and social benefits to local residents. The investigation
revealed that there were few differences in development between the affected area and the control area. Some benefits of the projects, such as
improved navigation and flood control, actually accrued to downstream
and existing urban areas. To overcome this maldistribution of social and
economic costs and benefits, Shanks recommends the use of advocacy resource development. Advocacy resource development assumes that the
local population has authority over its future and should have input into
federal water development. A limitation of advocacy planning is that it
emphasizes only local objectives and excludes national and regional considerations.

More research is needed to determine the social impacts of water resources development upon local communities. Such research should follow the community from the planning stage through the construction
stage and into the post-construction, use, and development stages. Impacts upon the local communities occur during each of these stages. To

optimize their choices, the communities and the implementing agency must know the available alternatives and the consequences of those alternatives.

Social Impact Analysis

As mentioned, past research has provided insight into the social, psychological, and economic effects upon individuals and families forced to relocate because of water resource development. With additional information on the effects of water resource development on communities, sociology can strengthen one of the more promising tools in making decisions about water resource development—social impact analysis.

Social impact analysis can provide public decision-makers involved in planning and implementing water resource development projects with information about possible consequences of alternative actions. The utility of social impact analysis comes from the fact that discrete bits of information are not considered in isolation. Rather, a number of variables are considered simultaneously within a model that is based upon theories of community and regional change and development (*28*). Although social impact analysis as a tool for decision-makers is far from being perfected, it is becoming increasingly more sophisticated. With additional research on the effects of water resource development on communities and the application of this information to specific models, social impact analysis will be enhanced through the improved predictive power of the models.

Conclusion

Water problems in urban areas always seem more evident and more critical than problems in rural areas. We hear about moratoria on new construction in Florida cities because of aquifer depletion and saltwater intrusion, about a fire on a river from industrial pollution in a northeastern metropolitan area, and about contamination by a petrochemical industry of an aquifer supplying water to a metropolitan area in southern Louisiana. Normally, the publicized water problems in rural areas are situations of too little water or too much water—drought or floods. And even floods make better news when they affect urban areas. Yet rural areas have problems of at least equal magnitude and intensity, and certainly they have a vested interest in the solution of these problems. Groundwater depletion in Nebraska and Kansas, surface water pollution in Ohio and southern Illinois, and erosion and sedimentation throughout the North Central Region are but a few examples of severe water problems in rural areas (*48*). Yet, because the population in rural areas is dispersed, the problems encountered by that population are also dispersed, i.e., the critical mass necessary to create broader recognition of the problems is difficult to achieve.

As we have seen, water quality issues frequently involve situations having salient impact on community processes, such as change, growth, conflict, and goal orientation. Accompanying increased demands upon the rural sector for the production of agricultural products and the provision of recreational amenities, in conjunction with constraints upon the ability to meet these demands, is an acceleration of water problems in rural areas. There are no quick, simple solutions to these problems.

Sociologists have a contribution to make, but they should not be viewed as alchemists who will gild this leaden situation. Rather, interdisciplinary research that combines technical, economic, political, and social factors offers the greatest potential benefit.

REFERENCES

1. Aiken, J. David, and Raymond J. Supalla. 1979. *Groundwater mining and western water rights law: The Nebraska experience.* South Dakota Law Review 24(3): 607-648.
2. Arnstein, Sherry R. 1969. *A ladder of citizen participation.* Journal of the American Institute of Planners 12 (July): 216-224.
3. Burdge, Rabel J. 1982. *Needs assessment surveys for decision-makers.* In Don A. Dillman and Daryl J. Hobbs [editors] *Rural Society in the U.S.: Issues for the 1980s.* Westview Press, Boulder, Colorado. pp. 273-283.
4. Burdge, Rabel J., and Paul Opryszek. 1981. *Coping with change: An interdisciplinary assessment of the Lake Shelbyville Reservoir.* Institute for Environmental Studies, University of Illinois, Urbana.
5. Christiansen, Douglas A., Andrew Morton, and Earl O. Heady. 1981. *The potential effect of increased water prices on U.S. agriculture.* Report 101. Center for Agriculture and Rural Development, Iowa State University, Ames.
6. Council for Agricultural Science and Technology. 1981. *Preserving agricultural land: Issues and policy alternatives.* Report No. 90. Ames, Iowa.
7. Clark, Roger N., John C. Hendee, and George H. Stankey. 1976. *CODINVOLVE: A tool for analyzing public input.* In John C. Pierce and Harvey R. Doerksen [editors] *Water Politics and Public Involvement.* Ann Arbor Science Publishers, Ann Arbor, Michigan. pp. 145-177.
8. Collette, W. Arden, Earl O. Heady, and Kenneth J. Nicol. 1976. *The impact of water rights and legal institutions on land and water use in 2000.* Report 70. Center for Agricultural and Rural Development, Iowa State University, Ames.
9. Davies, Race D., and Bruce A. Haines. 1978. *Some political-institutional factors affecting efforts to conserve water in Washington State.* Washington Water Research Center and Department of Political Science, Washington State University, Pullman.
10. Doerksen, Harvey R., and John C. Pierce. 1976. *Citizen influence in water policy decisions: Context, constraints and alternatives.* In John C. Pierce and Harvey R. Doerksen [editors] *Water Politics and Public Involvement.* Ann Arbor Science Publishers, Ann Arbor, Michigan. pp. 3-19.
11. Francis, Joe D. 1982. *Water.* In Don A. Dillman and Daryl J. Hobbs [editors] *Rural Society in the U.S.: Issues for the 1980s.* Westview Press, Boulder, Colorado. pp. 382-388.
12. Freeman, David M. 1977. *A social well-being framework for assessing resources management alternatives.* In Ved P. Nanda [editor] *Water Needs for the Future.* Westview Press, Boulder, Colorado. pp. 153-169.
13. Goldschmidt, Walter. 1978. *As you sow: Three studies in the social consequences of*

agribusiness. Allanheld, Osmun, Montclair, New Jersey.

14. Hogg, Thomas C. 1968. *Social and cultural impacts of water development.* In Emery N. Castle [editor] *People and Water.* Oregon Water Resources Research Institute, Oregon State University, Corvallis. pp. 11-23.

15. Hoiberg, E., P. Nowak, D. Albrecht, J. Bohlen, and G. Bultena. 1980. *Land use planning in Iowa: A study of farmers' attitudes toward planning issues.* Report No. 147. Department of Sociology and Anthropology, Iowa State University, Ames.

16. Huffman, James L. 1980. *Agriculture and the Columbia River: A legal and policy perspective.* Environmental Law 10: 281-314.

17. Jensen, Dallin W. 1977. *Some legal aspects of water resources management.* Public Administration Review 37: 456-462.

18. Johnson, Sue. 1974. *Recent sociological contributions to water resources management and development.* In L. Douglas James [editor] *Man and Water: The Social Sciences in Management of Water Resources.* University of Kentucky Press, Lexington. pp. 164-199.

19. Jones, Nancy. 1978. *Proposed rules for administering the acreage limitation of reclamation law.* Natural Resources Journal 18(3-4): 934-940.

20. Kansas Water Resources Research Institute. 1980. *Five year research plan, 1982-1987.* Kansas University, Lawrence.

21. Keith, Robert F., and George R. Francis. 1980. *Educating people to adjust to resource-constrained economies: An overview.* In *Resource-Constrained Economies: The North American Dilemma.* Soil Conservation Society of America, Ankeny, Iowa. pp. 271-277.

22. Kelley, Amy K. 1978. *Acreage and residency limitations in the Imperial Valley: A case study in national reclamation policy.* South Dakota Law Review 23(3): 621-661.

23. Klessig, Lowell L., and Victor L. Strite. 1980. *The ELF odyssey: National security versus environmental protection.* Westview Press, Boulder, Colorado.

24. Korsching, Peter F. 1982. *Behavior intentions as a technique for measuring commitment with community surveys.* Sociological Practice 4(1): 57-75.

25. Korsching, Peter F., and Peter J. Nowak. 1982. *Farmer acceptance of alternative conservation policies.* Agriculture and Environment 7: 1-14.

26. Lackmann, Christopher. 1977. *Reclamation Act of 1902: After 75 years 160 acres limitation held valid.* National Resources Journal 17(3,4): 673-678.

27. Lee, Motoko Y. 1981. *Mandatory or voluntary water conservation: A case study of Iowa communities during drought.* Journal of Soil and Water Conservation 36(4): 231-234.

28. Leistritz, Larry F., and Steven H. Murdock. 1981. *The socioeconomic impact of resource development: Methods for assessment.* Westview Press, Boulder, Colorado.

29. Leitko, Thomas A. 1977. *Issues, interests and power: Environmental politics in the community setting.* Ph.D. dissertation. University of Delaware, Newark.

30. Loehr, Raymond C. 1974. *Agricultural waste management: Problems, processes and approaches.* Academic Press, New York, New York.

31. Mazmanian, Daniel A. 1976. *Participatory democracy in a federal agency.* In John C. Pierce and Harvey R. Doerksen [editors] *Water Politics and Public Involvement.* Ann Arbor Science Publishers, Ann Arbor, Michigan. pp. 201-223.

32. Muckleston, Keith W., and Richard M. Highsmith, Jr. 1978. *Center pivot irrigation in the Columbia Basin of Washington and Oregon: Dynamics and implications.* Water Resources Bulletin 14(5): 1,121-1,128.

33. Nowak, Peter J. 1982. *Phase one final report of the selling of soil conservation: A test of the voluntary approach.* Department of Sociology, Iowa State University, Ames.

34. Oeltjen, Jarret C., and Loyd K. Fischer. 1978. *Allocation of rights to water: Preferences, priorities and the role of the market.* Nebraska Law Review 57(2): 243-282.

35. Office of Technology Assessment. 1982. *Impacts of technology on U.S. cropland and rangeland productivity.* Washington, D.C.

36. Radosevich, George E., and Melvin B. Sabey. 1977. *Water rights, eminent domain, and the public trust.* Water Resources Bulletin 13(4): 747-781.

37. Rickson, R. E., P. J. Tichenor, G. A. Donahue, and C. Olien. 1975. *Role of the scientist technician in water policy decisions at the community levels: A study in purposive communication.* Bulletin No. 79. Minnesota Water Resources Research Center, University of Minnesota, St. Paul.
38. Rickson, Roy E. 1977. *Dimensions of environmental management: Legitimation of government regulation by industrial managers.* Environment and Behavior 9: 15-40.
39. Rivkin/Carson, Inc. 1971. *Population growth in communities in relation to water resources policy.* National Technical Information Service, Springfield, Virginia.
40. Rogers, Everett M., and F. Floyd Shoemaker. 1971. *Communication of innovations: A cross cultural approach.* Free Press, New York, New York.
41. Rose, Douglas D. 1976. *Public opinion and water policy.* In John C. Pierce and Harvey R. Doerksen [editors] *Water Politics and Public Involvement.* Ann Arbor Science Publishers, Ann Arbor, Michigan. pp. 179-200.
42. Schneider, Anne L. 1976. *Measuring political responsiveness: A comparison of several alternative models.* In John C. Pierce and Harvey R. Doerksen [editors] *Water Politics and Public Involvement.* Ann Arbor Science Publishers, Ann Arbor, Michigan. pp. 87-115.
43. Shanks, Bernard. 1977. *Missouri River development policy and rural community development.* Water Resources Bulletin 13(2): 255-263.
44. Sloggett, Gordon. 1981. *Prospects for groundwater irrigation: Declining levels and rising energy costs.* Economic Report No. 478. Economic Research Service, U.S. Department of Agriculture, Washington, D.C.
45. Tregarthen, Timothy D. 1977. *The market for property rights in water.* In Ved P. Nanda [editor] *Water Needs for the Future.* Westview Press, Boulder, Colorado. pp. 139-151.
46. Trelease, Frank J. 1977. *Alternatives to appropriation law.* In Ved P. Nanda [editor] *Water Needs for the Future.* Westview Press, Boulder, Colorado. pp. 59-81.
47. U.S. Department of Agriculture. 1980. *Soil and Water Conservation Act: Appraisal 1980 (review draft).* Washington, D.C.
48. U.S. Water Resources Council. 1978. *The nation's water resources, Part II: Water management problem profiles.* Washington, D.C.
49. Walker, Lewis D. 1980. *An assessment of water resources in the United States, 1975-2000.* In *Resource-Constrained Economies: The North American Dilemma.* Soil Conservation Society of America, Ankeny, Iowa. pp. 73-95.
50. Warren, Roland. 1972. *The community in America.* Rand McNally, Chicago, Illinois.
51. Wengert, Norman. 1976. *Participation and the administrative process.* In John C. Pierce and Harvey R. Doerksen [editors] *Water Politics and Public Involvement.* Ann Arbor Science Publishers, Ann Arbor, Michigan. pp. 29-41.
52. White, Michael D. 1977. *Legal restraints and responses to the allocation and distribution of water.* In Ved P. Nanda [editor] *Water Needs for the Future.* Westview Press, Boulder, Colorado.
53. Whittlesey, Norman K. 1980. *Irrigation development in the Pacific Northwest: A mixed blessing.* Environmental Law 10: 315-329.
54. Willeke, Gene E. 1976. *Identification of publics in water resources planning.* In John C. Pierce and Harvey R. Doerksen [editors] *Water Politics and Public Involvement.* Ann Arbor Science Publishers, Ann Arbor, Michigan. pp. 43-62.
55. Willeke, Gene E. 1978. *Assessing the social effects of water quality management programs.* Report No. 03-78. Georgia Environmental Resources Center, Georgia Institute of Technology, Atlanta.
56. Zaltman, Gerald, and Robert Duncan. 1977. *Strategies for planned change.* John Wiley & Sons, New York, New York.

IV
Water Research Organization
and Funding Alternatives

13

Options for Organizing Water-Related Research and Funding in Agricultural Experiment Stations

Raymond J. Supalla

Despite the substantial contributions of the agricultural research establishment over the past several decades, opportunities for future research payoffs have not diminished. Scientific progress appears to enhance the opportunities for further research payoffs, rather than reducing the need for research. This characteristic of science, when viewed in the context of a rapidly changing socioeconomic environment, creates a tremendous challenge for the agricultural experiment stations in the North Central Region. If tomorrow's research needs and opportunities are to be met, we must find better ways of efficiently administering and financing the required research.

The Current Situation

In reviewing the current financial and administrative situation, my intent is to bring about an understanding of the current research environment, with emphasis on those aspects that might be modified to enhance the ability of experiment stations to meet water research needs.

Financial Setting

Historically, water resources research has been funded well (*1, 4*). This was especially true after passage of the Water Research and Education Act of 1965, which established the water resources research institutes in each state. With the advent of Reaganomics in the 1980s, however, the

225

situation reversed. Economic recession and a reallocation of governmental priorities has created a relatively bleak financial picture.

Since 1980, federal support for water resources research has diminished significantly. This is reflected in the elimination of the U.S. Department of the Interior's Office of Water Research and Technology (OWRT); in the reduction of the research roles of the U.S. Environmental Protection Agency (EPA) and the U.S. Department of Energy (DOE); and in the steady erosion of the budget for the Cooperative States Research Service (CSRS) in real terms. The loss of OWRT in 1982 has meant at least a temporary reduction in federal research support of about $15 million; the CSRS budget has declined slightly in real terms over the past five years; and the extensive applied research programs of EPA and DOE have been all but eliminated.

Current state agricultural research support within the North Central Region is unfavorable also. State appropriations to experiment stations within the region increased 27 percent from 1977 to 1980, while inflation exceeded 36 percent (2, 3). Moreover, the onset of recession has surely made the situation worse, but specific data regarding the magnitude of more current appropriations are not readily available.

This trend toward more stringent research funding has affected research productivity in several ways. The funding situation has resulted in real reductions in operating dollars within most experiment stations. This has reduced the flexibility that scientists have in developing research programs, making it increasingly difficult to pursue high-priority problems with large potential payoffs. It has also meant that scientists are spending more time soliciting dollars and administering research projects and less time doing creative research.

There is one final impact on research productivity. As the funding noose tightens, it becomes increasingly difficult to maintain a critical mass of scientists in specialized problem areas. This prevents the effective pursuit of some types of research and increases the likelihood that more "blind alleys" of investigation will be pursued.

Current Administrative Practices

In addressing how to pursue emerging water related research needs more effectively, it is important to review some current administrative practices that may need to be reassessed. These include the organizational structure of experiment stations, the budget allocation process, and administrative staffing.

The basic administrative units of most agricultural experiment stations consist of academic disciplines. Although this structure was certainly appropriate in the past and may remain so in the future, an increasing number of emerging research problems appears to require a multidisciplinary approach. Even though multidisciplinary work can be and often is pursued within a discipline-oriented structure, the question is whether a different administrative structure might work better.

Another widely used administrative practice that may be inconsistent with emerging water-related research needs concerns the budget allocation process. Experiment station research budgets frequently are allocated to departments, then suballocated to projects. Again, this makes it difficult to pursue multidisciplinary problems because an individual scientist cannot retain the services of someone in another discipline without reducing his home department's research funding.

Still another important aspect of the current budgeting process concerns the extensive use of long-term financial commitments to departments and individuals. Although long-term commitments provide stability and research program continuity, they also limit flexibility. In the past, research program growth has permitted the research establishment to have both stability and flexibility. In a era of negative real growth, however, there may be a need to reassess historic policies.

A final administrative factor that may impinge on the ability of experiment stations to meet emerging research needs concerns the issue of administrative staffing. With reduced research support in real terms and growing research needs, creative research program administration becomes more important. The common response appears to be one of saving dollars by cutting administrative staffing. Perhaps this is penny wise and pound foolish? Use of skeleton administrative staffs leads to minimal program planning and results in scientists receiving relatively little assistance in pursuing grants and contracts.

Criteria for Evaluating Options

Before proceeding to suggest ways of increasing the ability of agricultural experiment stations to meet emerging water research needs, criteria are needed for choosing among the options. Although the suggested criteria are neither exhaustive nor universally agreed upon, they do set the stage for thinking about alternatives.

The criteria used here to evaluate administrative and financing options include the following:

1. The administrative structure must allow for both disciplinary and multidisciplinary programs.

2. The incentive/rewards structure must encourage desired activity, yet be perceived as equitable by all scientists.

3. Research programs must have general public value, i.e., client-specific projects need to be minimized.

4. Funding procedures must balance funding sources (willingness to pay) with research priorities, as determined by scientists.

5. Funding procedures and programs must preserve academic honesty.

6. Research program specifications should emphasize efficiency relative to general public needs, i.e., maximize research payoffs per dollar expended, even when this comes at the expense of equity relative to specific client groups.

Ideas for Administrative Change

There appear to be many administrative practices that could be implemented to enhance the ability of experiment stations to meet water research needs, although none of the options are without trade-offs. The intent here is to raise some selected ideas and to assess their advantages and disadvantages relative to the criteria above.

One idea that merits consideration is to designate a portion of experiment station budgets for exclusive use on multidisciplinary research projects. This budget item would be directly administered by the experiment station and would not become part of departmental allocations. Such a program would provide incentives for multidisciplinary cooperation because scientists could get access to the resources only through participation in a research team. It would also provide for research management via purse-string incentives. In this context, a scientist who becomes project leader would have financial leverage to induce contributions from other team members, whereas no such leverage is present when the team consists merely of cooperators using their respective departmental budgets.

The major disadvantage of such an action is that department heads would control a smaller proportion of the total budget. This would reduce departmental flexibility and possibly fragment the evaluation and management of individual faculty programs.

A second idea for administrative change concerns providing salary supplements to scientists who secure grant funding. Although this is now done indirectly in most cases, in that grant funding success becomes part of performance evaluation, it is often not apparent that such rewards are present. To make the rewards for securing research grants direct and explicit would probably result in incentive-induced increases in total research support. It may also mean that scientists are more equitably rewarded, leading to higher morale and greater productivity. Perhaps more importantly, such a program would discourage the conduct of research outside university channels. In considering this aspect, it is important to note that when faculty salaries fail to keep pace with inflation the nearly inevitable result is increased consulting activity, including the conduct of university-type research in privately established firms composed of faculty members.

A potentially critical disadvantage of a grant incentive program is that the financial reward structure may distort a scientist's view of research needs. This is unlikely to have a negative impact, however, if one assumes that grant funds tend to be available only for important problems that could not otherwise be researched. A second, closely related disadvantage is that grant incentives may lead to too much emphasis on problem solving rather than on basic research. This issue may be especially critical for water resources research; historically, it has been much easier to secure grants for applied rather than basic research.

A third administrative idea is to expand the use of research centers and institutes, but with carefully defined roles. Before considering this issue, however, it is important to differentiate between a research center and a research institute. In the absence of widely accepted definitions, a research center is defined here as an administrative entity responsible for stimulating, coordinating, and funding research in a broad general area, such as water resources. In contrast, a research institute is defined as an administrative entity responsible for actually conducting a focused research program that cuts across disciplinary lines.

Research centers can contribute to meeting research needs by assisting scientists in the identification and pursuit of research funding, by assisting in the distribution of research results to users, by assisting in the communication of research needs from clients to users, and by providing research administration services, which, among other things, would facilitate pursuit of multidisciplinary research projects.

A research center is appropriate when what is primarily desired is research funding assistance and broad program coordination. In contrast, a research institute is the appropriate entity when what is desired is an administrative home for a rather narrowly focused, multidisciplinary research program. In terms of specific functions, research institutes simplify research administration when people from different disciplines are involved and provide a home for equipment and support personnel used in multidisciplinary efforts.

The disadvantages or problems associated with research centers and institutes are not insurmountable. The major difficulty rests in delineating their respective roles relative to other administrative entities. When roles are not carefully delineated, empire building can occur, leading to fragmentation of effort. Another problem is that affiliated staff may find themselves alienated from the disciplines, especially in the case of focused research institutes, where staff may answer directly to institute directors rather than to department heads.

A final administrative idea concerns the use of increased state specialization in research. As budgets tighten and needs increase, it is becoming increasingly appropriate for state agricultural experiment stations to entertain cooperative specialization. For example, one state might elect to focus on irrigation engineering while another focuses on basic hydrology.

The major advantage of specialization is increased efficiency in terms of equipment, facilities, and support personnel. This is especially true for those types of research where extensive capital expenditures are required. Specialization makes it possible for the same number of scientists to share better facilities, or the same quality facilities, at lower cost.

A second important advantage of state specialization in research is that it would be easier to maintain a critical mass of scientists in a single location. Although it is possible to pool ideas across state lines through symposia, telephone communications, and the like, many research endeavors require day-to-day interaction of a type that only can be achieved when the personnel involved are at the same location.

Increased state specialization appears desirable on the surface, but there are some important trade-offs to consider. First, state specialization makes communication among complementary groups difficult. In cases where the research involved does not require extensive communication with personnel outside the immediate specialized group, state specialization would probably work well. On the other hand, for research problems where many different skills are required an emphasis on state specialization may reduce effectiveness because of communication problems. For example, if one state specialized in irrigation engineering while another specialized in hydrology, the situation may work well until the irrigation engineers located in state X get involved in a problem that requires a close working relationship with hydrologists in state Y.

A second disadvantage or problem involves the difficult decisions regarding who will specialize in what for how long. Until such time as there exists a research czar with authority over all universities, it is unlikely that specialization will occur, except in those rare instances where there is an opportunity for two or more states to gain equally from some arrangement.

A final problem associated with state specialization concerns relationships between teaching and research. Although there may be several instances where specialization is appropriate from a research perspective, the teaching dimension may make such action undesirable. This might be especially true at the graduate level, where effective graduate teaching requires an active research program. Although the prospective graduate student interested in a particular specialty would clearly go to the state with strength in his or her area of interest, one may have some difficulty providing adequate training in support areas if the appropriate subspecialty is located elsewhere.

Ideas for Financing Water Resources Research

There are at least four ways in which additional water resources research funds can be acquired: (1) encourage additional general fund support at the state and federal levels, (2) expand the use of user fees, (3) expand the use of special taxes or checkoffs, and (4) establish specialized grant and contract offices within agricultural experiment stations.

Within each of these categories there are several things that could be done. There are also some important trade-offs to consider before electing to implement any given idea.

All experiment stations are engaged in a continuous process of encouraging general fund support, and the usual processes can and should continue. At the same time, there appear to be some unexploited opportunities. The first would be to engage in additional research directed at evaluating and documenting returns to research and education. The proceedings of this symposium constitutes one form of evidence of research payoff that could be used effectively. This approach could be taken one step further by commissioning studies that would measure actual rates of return to specified types of research programs.

A second funding activity that appears to merit greater emphasis than it receives in most states is an aggressive public relations program designed to broaden clientele-group affiliations. A cursory review of the literature reveals that those who profit most from agricultural research are oftentimes not the groups most courted by experiment station administrators (5). The group that gains the most from agricultural research consists of consumers of food and fiber products, yet it is the rare experiment station director who cultivates relationships with consumer groups.

When one considers water resources research specifically, it becomes evident that potential support groups are especially numerous. Research that leads to more efficient uses of water contributes to fish and wildlife, environmental quality, lower food costs, more tourism, increased net farm income, reduced transportation costs, and reduced electricity costs. Despite these broad payoffs, it is usually only the conventional agricultural interests that are courted as political supporters of funding needs. Clearly, there are untapped opportunities for creating additional support for research funding within agricultural experiment stations.

A third method of increasing general fund support for research consists of crying "uncle" more often. One can surely argue that during hard financial times the public reasonably expects a sharing of revenue shortfalls. Thus, it may be politically wise to not scream too loudly about revenue needs. Conversely, it can be argued that research and education constitute an investment that can only be reduced at extremely high long-run costs to society. If one accepts this latter argument, it follows that the responsible approach is to insist that society cannot afford real research funding reductions. Convincing presentation of such an argument to the public will almost certainly lead to increased general fund support.

The second general category of ideas for increasing research funding involves expanding the use of user fees. In this instance, the specific possibilities are numerous. One could go as far as duplicating the operations of a private consulting firm. Given the philosophy and role of land grant institutions, however, such extremism is inappropriate. It nevertheless appears that some user fees could be instituted without violating the land grant charter.

The most obvious user fee option available to experiment stations consists of publication charges, and many universities have begun to make wider use of such charges in recent years. Such a program is often of little value in generating research funds, however, because the largest number of publications are distributed by the Cooperative Extension Service, with the revenues accruing to extension, not research. It may be time to rethink this approach and recognize that many, if not most, extension publications are reporting on work done by experiment stations. Given the contributions of experiment station scientists to the content of extension publications, it seems reasonable to propose that the revenue generated through publication sales be shared equitably.

In considering publication charges it is important to recognize that it may be necessary to charge an amount greater than publication costs. Although the efficiency conditions associated with marginal-cost pricing call for recovering publication costs only, it may be appropriate to violate this condition in order to generate the long-term returns associated with additional research. At the same time it must be recognized that publication charges cannot exceed printing costs by much before few will be purchased and less, not more, revenue will be generated.

Another user fee issue concerns charges for client-specific work. Although agricultural experiment stations are public institutions and client-specific work is appropriately discouraged, some inevitably occurs. Research funding could be increased significantly in many cases if all client-specific work were performed for fees that at least recover full costs. The actual fees charged should correspond to market values, including "profits" when available.

The third general category of funding ideas concerns various types of special taxes or check-offs. Although university research historically has been supported by general revenues, a reasonable argument can be made for funding research programs in part from special taxes. An obvious example in the case of water resources research involves water use charges. Because the most direct beneficiaries of research directed at improving water use efficiency are water consumers, why not implement a charge per unit of water consumed, with the proceeds to be used at least in part for research programs? Following the same philosophy, one might go a step further and place surcharges on boat and fishing licenses in order to get a contribution from those who benefit directly, but who do not consume water.

Commodity check-offs are another type of special tax that one might use for general experiment station support. Commodity check-offs are now being used extensively by commodity groups in some states to generate revenue for market promotion activities and, occasionally, for research. It would be a small step for experiment stations to place greater emphasis on this source of funds. The ideal situation would be to have a portion of the check-off earmarked for agricultural research, thus making it a continuing source of "hard" research support.

Another special tax alternative would involve placing the tax at a different stage in the production process. Because it is consumers who ultimately receive much of the benefit from agricultural research, there would be justification for placing an excise tax on food and fiber products, with the proceeds designated for experiment station research. One problem with this approach, however, is that many would consider such a tax unfair to low-income people who spend a larger proportion of their total income for food and fiber products. This problem could be solved at least in part through the use of low income tax credits. In sum, it is an attractive option for funding research because a small tax percentagewise would generate large amounts of revenue. It is also attractive because such a tax would be relatively consistent with the principle of taxing in proportion to benefits received.

The final category of research funding alternatives concerns the establishment of specialized grant and contract offices within agricultural experiment stations. All major universities have grant and contract offices, and since 1965, all land grant universities within the North Central Region have had water resources research centers as well. These entities provide an important array of administrative services, but their responsibilities are broad, and thus, they often fall short of providing the detailed assistance necessary to maximize effectiveness in grantsmanship competition. What appears to be needed for water resources research funding are water research centers at the experiment station level and/or an expanded grants and contracts staff located within each of the experiment stations. Such entities would have a narrow enough charge to facilitate provision of the detailed assistance necessary for effective grantsmanship.

A water resources center or a water resources grants and contracts officer within an experiment station would appropriately be responsible for monitoring research funding opportunities within state and federal government, cultivating relationships with private foundations and other potential private donors, assisting scientists in the preparation of grant proposals, and assisting scientists in grant project administration. With the recent demise of OWRT, along with a general tightening of available research funds, these grantsmanship services are especially critical to acquiring the necessary funds for meeting emerging water resources research needs.

Recommendations

In view of the emerging water resources research needs and the generally bleak research funding situation, it is imperative that experiment stations aggressively pursue all reasonable alternatives for improving research program efficiencies and enhancing the support base.

The administrative adjustments that appear to have the most merit include the following:

1. Experiment station administration of multidisciplinary programs, culminating in more extensive use of research institutes.

2. Establishment of policies that provide for salary supplements to scientists who secure grant funding.

3. Development of additional cooperative relationships among states with similar research needs, culminating in a higher degree of research program specialization, with associated improvements in efficiency.

Securing adequate funding for emerging water resources research needs will require expanded efforts to increase state and federal support, as well as increased reliance on nontraditional funding sources, with particular emphasis on user charges, special taxes, and aggressive grantsmanship. The specific options that appear to have the greatest potential include additional programs to document publicly the value of research, additional efforts to secure the support of nonagricultural political groups who profit from water-related research, publication charges, higher fees for client-specific research, and a water use tax.

The agricultural experiment stations of tomorrow will play a major role in providing for social needs through the efficient management of natural resources. To meet this responsibility, however, it will be necessary to abandon many traditions. But the challenge will be met and tomorrow's experiment stations will have less of a disciplinary emphasis, operate with proportionately less dependence upon general tax support, and have a more focused or specialized research program.

REFERENCES

1. Barnes, Carlton B. 1955. *What research is doing on problems in water in agriculture.* In *Yearbook of Agriculture.* U.S. Department of Agriculture, Washington, D.C.
2. Cooperative State Research Service. 1979. *Inventory of agricultural research FY 1977. Volume II.* U.S. Department of Agriculture, Washington, D.C.
3. Cooperative State Research Service. 1982. *Inventory of agricultural research FY 1980. Volume II.* U.S. Department of Agriculture, Washington, D.C.
4. Joint Task Force, U.S. Department of Agriculture and State Universities and Land Grant Colleges. 1969. *Program of research for water and watersheds.* U.S. Department of Agriculture, Washington, D.C.
5. North Central Committee 148. 1981. *Evaluation of agricultural research.* Miscellaneous publication 8. University of Minnesota, St. Paul.

Discussion

Roy M. Gray

The agricultural experiment stations, in conjunction with the water resource institutes, continue to do an outstanding job of water resource research. As Supalla indicates, however, public resources are likely to be more difficult to obtain because of budget tightening by federal and state governments. In the federal sector, concern over the budget deficit probably will mean that the only research efforts to be funded will be those perceived as addressing the most urgent needs. Research institutions will have to do more with less, or simply do less. *Efficiency* in the use of funds and personnel will be critical to the success of research by the experiment stations and water resource institutes.

Another key, in my judgment, will be the ability of individual researchers to produce results that can be applied on the land. Consultation and cooperation among researchers who generate new information and agencies, such as the Extension Service and the Soil Conservation Service, that have the responsibility to transfer information to the farmer are even more essential now than in the past. This dialogue among researchers and agencies will ensure that needs of the highest priority are addressed and that the research results are used. This is not meant as an argument for applied research as opposed to basic research. Both are important, but funds will be even more difficult to obtain unless a relatively short-term payoff can be demonstrated.

Supalla analyzes well the situation facing water resource research efforts. His conclusions should be of particular interest to administrators of water research programs. A couple of his recommendations, however, may not be consistent. He recommends that "research program specifications should emphasize efficiency relative to general public needs... even when this comes at the expense of equity relative to specific client groups." But he also recommends establishing "policies that provide for salary supplements to scientists who secure grant funding." While these two statements are not *necessarily* inconsistent, grantsmanship carries with it a responsibility to provide a product that is of specific use and special interest to the grantor. The results may or may not meet the highest public need.

Supalla also seems to argue for the establishment of special administrative units to facilitate obtaining outside funding and conducting multidisciplinary research. Creating a separate administrative entity, such as a research center or institute, is bound to result in some administrative cost above that already incurred by the existing administrative unit, the experiment station. The water resource institutes can directly fund work within the colleges of agriculture, engineering, liberal arts, and others not traditionally funded by the agricultural experiment station. If the the experiment station cannot now provide this direct funding when such areas of expertise are needed to solve a problem; the needed personnel must be added to the station staff. It is difficult to see why this barrier to multidisciplinary work cannot be overcome. Particularly in times of budgetary restraint, any institution, whether a federal agency, university, or private company, will look skeptically at proposals to create a new administrative entity.

Discussion

R. J. Hildreth

The major problem in funding emerging water research needs is the erosion of funding for all research, not just water research. This erosion makes increases in funding of water research by agricultural agencies difficult.

What are the elements of the erosion? First is the recent erosion in federal and state support in real terms, as Supalla points out. Second is the possible reduction in funds from federal sources for research in nominal terms, i.e., absolute cuts.

What leads me to suggest possible absolute cuts in the face of strong support for federal funding of agricultural research by Secretary John Block? Five factors:

1. The projected deficits ($100 billion-$200 billion) in the federal budget for many years into the future. These large deficits will put great pressure on federal funding for all research programs.

2. The growing consensus within the Reagan administration that the major, perhaps only, role of federal support for research is support for basic research. In this framework, states and firms should support applied research. A great deal of useful agricultural research is applied.

3. The failure of the state and federal agricultural research system to present to the political system an image of a coordinated program instead of an image of a duplicative, uncoordinated series of activities. The image of duplicative, uncoordinated programs is enchanced by keen competition for funds among regions, commodities, subject matter areas, and agencies, e.g., state versus federal.

4. A perceived decline in the legitimacy of research by many publics and the political system. Many people appear to think that agricultural research is an entitlement program for researchers rather than an investment by society.

5. A lack of understanding by consumers and the political system that the major benefits of agricultural research go to consumers, albeit a small amount per family. A recent Office of Technology Assessment study suggested that the benefit per family is on the order of $25 a year; multiplied by the number of families, however, the benefit is a very large number (2).

Evidence is strong that the research system has been a good public investment, returning a yield of about 50 percent to the public dollar (1). This positive performance is overwhelmed by the above five factors.

What are the implications for water research administration and funding, especially funding from the federal level? There are four:

1. All levels of the agricultural research establishment, scientists and administrators, need to realize the major issue may well be survival. The appropriate behavior for survival is not to give up, but to realize that survival behavior is quite different than business as usual.

2. The justification for funding water research should not come at the expense of other subject matter in conversations with the political system.

3. There is a need to present to the political system a coordinated program of research. This will require agreement on priorities for agricultural research.

4. Ways to increase efficiency in the performance of agricultural research need to be found.

Some examples of increasing efficiency include better research problem definition. Administrators and scientists need to find ways to better use scarce time and equipment. Specialization is often a way to increase efficiency. In spite of possible reductions in funding, specialization needs to continue. Coordinated research between and among some state agricultural experiment stations and federal research agencies may assist in specialization. By working together in formal or informal regional research efforts, state agricultural experiment stations and federal research agencies will be able to achieve some specialization. Agricultural research has made significant strides in increasing efficiency in agricultural production and marketing. Perhaps formal research on alternative ways for improving efficiency in research itself could yield positive results. For example, the use of models could be used to simulate field trials in engineering and agronomy. Such simulations might lead to finding the crucial experiments to be undertaken by field trials.

The agricultural research system has performed magnificently over its history. As it faces the possibilities of reduced funding, it can draw upon resources deep within the system to face this challenge and opportunity.

REFERENCES

1. Hildreth, R. J. 1982. *The agricultural research establishment in transition.* Proceedings, Academy of Political Science 34(3): 243.
2. White, Fred C., B. R. Eddleman, and J. C. Purcell. 1982. *Nature and flow of benefits from ag-food research.* In *An Assessment of the United States Food and Agricultural Research System.* Office of Technology Assessment, Washington, D.C.

Index